T0185631

The First Cell

Ulrich C. Schreiber • Christian Mayer

The First Cell

The Mystery Surrounding the Beginning of Life

 Springer

Ulrich C. Schreiber
Faculty for Biology
University of Duisburg-Essen
Essen, North Rhine-Westphalia, Germany

Christian Mayer
Faculty for Chemistry
University of Duisburg-Essen
Essen, North Rhine-Westphalia, Germany

ISBN 978-3-030-45383-1   ISBN 978-3-030-45381-7   (eBook)
https://doi.org/10.1007/978-3-030-45381-7

This Springer imprint is published by the registered company Springer Nature Switzerland AG.
The registered company address is: Gewerbestrasse 11, 6330 Cham, Switzerland

Life is a storm (in a teacup) in the stream of entropy
Ulrich C. Schreiber, February 11, 2019

For Karin, Hanna, and Sebastian

Preface to the English Edition

The following book tries to offer a comprehensive model for the formation of the first living cell. Its ideas have been formed during cooperation between the two authors over many years. The original edition, appearing as a non-fiction book in German, described the complex topic from a personal perspective of one of the authors (U.C. Schreiber).

The idea to write a non-fiction book on the topic of the origin of life was born on the results of laboratory experiments which prove the occurrence of a chemical evolution under environmental conditions. These experiments opened an opportunity to follow the development from simple organic molecules up to complex structures with particular functions. Based on this stage during the completion of the first edition, it was planned to extend the contents of this book with new developments and new ideas. This led to this second English version in which C. Mayer acts as a coauthor. C. Mayer essentially developed and performed the key laboratory experiments and lately opened a fundamental discussion on a new perspective of the definition of life. The latter is now included in this book as a new chapter (Chap. 9) on life in the context of order and complexity.

Preface

It's risky starting a project about the origin of life. In the search for approaches and solutions, hurdles quickly become apparent that appear insurmountable—above all those concerning your own knowledge. The origin of life is not a research object that in its fullness can only be assigned to biology, chemistry, or biochemistry. The sum of all questions that arise in the search for answers concerns a large number of scientific disciplines and, as it quickly became apparent, especially those of physical chemistry and geology. Perhaps this is also the reason why the answers that science has offered so far have been so unconvincing. Often a broad-based research group is lacking, which is absolutely necessary for dealing with all the aspects involved. All this quickly became clear once the idea of pursuing a completely new approach to researching the formation of organic molecules in the continental crust's fracture zones, and ultimately of life, was born in 2003 and 2004. It was Oliver Locker-Grütjen, Head of the Science Support Center at the University of Duisburg-Essen, whose attention I first drew to this consideration. We quickly agreed that specialists from all the natural sciences were needed to have any chance at all of gaining new insights into this question or finding answers to it. Embarking on such a venture on our own was unthinkable. Oliver Locker-Grütjen knew numerous colleagues from a variety of departments at the university and was able to assess who might be willing to take an unconventional approach to this topic. After a short while, more than ten professors came together who were so interested in the question of the origin of life that they were willing, at intervals of several months, to take part in privately organized evening meetings despite having full schedules, and thus, the Essen Origin of Life research group was born.

A lot has happened since then. This book presents the current state of research and—let us anticipate this much—shows, on the basis of a hypothetical model, a completely new path to the formation of the first divisible cell—and this under conditions that are realistic and to some extent still occur in the same environment today. It makes no claim to being complete or exclusive. What it can do, however, is help to develop an understanding of the processes required that lead to new experiments and a sound approach to this extremely complex matter. Because this much is certain: besides the beginnings of the universe since the beginning of the scientific age, no scientific question has remained as unresolved as the one concerning the origin of life.

Now it can commence.

Essen, North Rhine-Westphalia, Germany   Ulrich C. Schreiber
March 2019

Acknowledgments

A non-fiction book like this one that addresses such an extensive range of themes has a long history in its formation. In addition to research in the laboratory and field, it is based on numerous meetings and discussions with colleagues, at the university, at conferences, and in the private sector. I would like to express my sincere thanks to all those who gave information, made corrections, and provided support in the drafting of this non-fiction book, as well as all those who have made contributions to discussions and provided assistance over the past few years. Listed below, in alphabetical order, you will find the many colleagues and other harbingers of information who have provided constructive commentary on this subject over the years: Prof. Peter Bayer, Prof. Steven A. Benner (Florida), Prof. Volker Buck, Dr. Maria Davila Garvin, Prof. Gerald Dyker, Prof. Matthias Epple, Prof. Hans-Curt Flemming, Prof. Daniel Hoffmann, Prof. Gerhard Jentzsch, Prof. Frank Keppler, Prof. Ute Klammer, Prof. Ralf Littke, Dr. Oliver Locker-Grütjen, Prof. Franco Pirajno (Perth), Prof. Agemar Siehl, Prof. Torsten Schmidt, Prof. Oliver J. Schmitz, Prof. Heinfried Schöler, Prof. Jörg Schröder, Prof. Bernd Sures, and Dr. Jonathan Williams. I would like to express my special thanks to the management team at Duisburg-Essen University, who showed great commitment in their support for the project.

The translation was kindly sponsored by SITEC-Sieber Engineering AG, Switzerland.

Contents

1

Introduction

Abstract Nothing is as complex as life. And although we now know quite a lot about it, one crucial question has still remained unanswered since time immemorial: how did life come into being? Did it happen according to scientific laws, here on earth, elsewhere, or by the hand of a supernatural creator? This question has moved mankind for thousands of years. The answers provided by science have so far been unsatisfactory.

A group of scientists from the University of Duisburg-Essen has tackled this question of what the biggest unsolved problem in science is perhaps and may be able to make a decisive contribution to solving it. They are searching for the beginning, the very first cell from which all other cells originate. Their quest is about the definition of life, entropy, energy, information storage, and how everything began.

An overview of the conditions on earth after its genesis, the general conditions for life, and the position that scientific research takes in the history of mankind provides an introduction to the topic, which is one of the most exciting in scientific research.

1.1 The Origins of Life: What Makes This Question So Important to Us?

Can't we as human beings simply accept that what was formed a long time ago exists today without knowing exactly how and why? Well, the simple answer is no, we can't. The development of human beings and hence the development

© Springer Nature Switzerland AG 2020
U. C. Schreiber, C. Mayer, *The First Cell*, https://doi.org/10.1007/978-3-030-45381-7_1

of an abstract thinking organ, our brain, inevitably lead to questions about everything that happens in the environment of our brain. This has always been the case since a certain point in development. Questions exist that could be answered instinctively. The most obvious one is why does game recognize that a hunter is approaching shortly after they move closer? While in a similar situation under the same cover and at the same distance, it continues grazing without a care as is an easy kill. Experience provided the answer to this dilemma, which, after many repeated scenarios, demonstrated that wind direction plays the decisive role. Other questions about what causes lightning and thunder or rainbows, illness, death, and a lot more besides were unanswerable and were simply accepted as a given. They found their place in the realm of the uncontrollable, the divine, something superior to man. This represented a very successful method of reducing the burden on the psyche with regard to questions that could not be answered at that time.

The approach to answering such questions and many more besides changed with the establishment in human thinking of scientific principles. A coherent answer required proof that was reproducible and universal. A statement about the gravitational pull of the earth had to be valid on every continent or also on ships in the case of the oceans. It goes without question for us today that objects everywhere on the earth accelerate in free fall toward the center of the earth. The physical law associated with this, which Newton formulated in the second half of the seventeenth century, is known as the law of gravitation and was supplemented by Einstein's general theory of relativity in the twentieth century. These scientific laws mean that we know that masses attract each other, everywhere, throughout the whole universe. No salvationist or conspiracy theorist, no matter how charismatic, can deceive the general public into thinking today that this is not the case on the moon or any other planet.

This scientific way of thinking meant that things that had long been accepted as a given were gradually removed away from the realm of the divine. Thanks to the natural sciences, we only have a few things left in our world of thought whose explanation is still considered by some to be the creation of God. One aspect here is the Big Bang, which established itself as an explanatory model for the origin of the universe a few decades ago and has special character. And it is only thanks to physics that this aspect found its way onto the agenda at all. It opens up a dilemma based on the following fact: the explanation model for the Big Bang is the result of a consistent physical examination of the processes that took place after the Big Bang, postulated up to the present day, without any divine influence. Some scientists even take the difficulties involved in physically describing the actual start of the Big Bang as a reason to claim it as being God-given, since there is still no clear explanation for this. In the meantime, however, alternative ideas have been developed to explain the emergence of the universe. These ideas involve an infinitely vibrating expansion and contraction without the need for a single Big Bang as the initial stage [1, 2]. This theory has not made things any easier.

The second aspect which remains unanswered is the question posed at the beginning of this book on how life on earth originated. This question has probably moved humanity since we had the ability to ask questions. Directly connected with this are the "why?" and the "where to?" and the reflection on the meaning of life in general. In a world where intelligent thinking beings exist, but where scientific principles still remain unknown, solutions to answering major questions like this have to be found in other ways. From the beginning, the solution was called religion. It gave and still gives answers to topics that no one can understand through their simple experiences of the everyday world. The truthfulness or reproducibility of the statements is of no import here. What is important is to calm your own insecurities and fears, which inevitably arise when thinking in these "incredible" spheres.

1.2 What Exactly Is Life?

In addition to providing gains in fundamental knowledge, only natural science has contributed to depicting the complexity of life and has raised a wealth of seemingly unanswerable questions. It showed that life, which is so natural in its existence for us, is surprisingly difficult to define.

Don't we just have to take a look around us to see what life is? No, because it's not as simple as that. In science, an all-encompassing definition still does not exist that explains life and hence the starting point of life as well. This is a big difference to chemistry and physics, where, for example, theories exist to explain matter or forces that act. That said, we can at least specify criteria or characteristics that are key features of life and are accepted by all natural science disciplines. These are inevitably the physicochemical characteristics that make up a living biological system. And here you can already see that it should be a system that corresponds to our knowledge of biology. Some of the characteristics can also certainly occur in nonbiological systems; the combination and simultaneity sharpen the definition of a description of life.

At the top of the list of criteria is the existence of at least one cell, a compartment enclosed by a cell membrane. This is where the biochemical reactions take place that prevent the cell from dying or, in other words, ensure that it stays alive. Biochemical reactions require information storage, a metabolism for absorbing energy and for exchanging molecules from the environment, and catalysts for efficient chemical reaction chains. With a precisely tuned regulation, the interaction of all the components leads to the reproduction of cell components, and to growth, and the proliferation of the cell through its division. Added to this is the ability to adapt to changing environmental conditions and to develop into more complex molecular groups.

A small note in the margin in relation: what if we put all the functioning cell components of trillions of cells (except their cell membranes) in a large vessel or an almost closed hole in the earth and supply them with energy and add and remove the necessary molecules? All the processes that would otherwise take place in a cell (but without cell membrane-related reactions) would continue in this vessel. The products multiplied thus could enter other space through flow paths and multiply overall. Would we call this system life? We could simply take the position that we do not need this to any thought, because the molecular cocktail cannot multiply, of course. But the idea is still important, because the end of this book gives us model ideas about the beginnings of organic chemistry up to the formation of vesicles and cells, which come close to such a situation and therefore need to be isolated.

As early as in 1980, the theoretical physicist Gerald Feinberg and chemist Robert Shapiro tried to make the principle of life universally applicable to other possible forms of life in space. They concluded that life originates from interactions between free energy and matter. In this way, matter is able to achieve a greater order within the common system [3].

Today, we can imagine a colony of robots that extract the raw materials from which they are made on their own, process them into components, and use them to reproduce themselves. They would be controlled by a computer, and each would have outer shell and solar cells on their body to generate energy. The metabolism would be defined by the entire colony, and artificial intelligence would ensure adaptation to changing environmental conditions. Most of the components could even consist of organic chemical components. In contrast to biological life, which developed itself on a physicochemical basis, a colony of robots would be the result of human creation. Would we attribute the facet of life to this colony?

It becomes apparent that borderline areas exist that require longer discussion. From a certain point onward, the step toward life as we know it was taken. In the period prior to this, a transition from purely physicochemical to information-driven organic molecule formation must have taken place. This important period is narrowed down further in Sect. 8.3. Two further examples should show you how difficult it is to clearly describe life in just a few words. A group of experts around the chemist Gerald Joyce coined the definition: "Life is a self-sustaining chemical system with the capacity for Darwinian evolution" [4]. The US space agency NASA also uses it as a working definition. Stuart Kauffman, a US American theoretical biologist, on the other hand, focuses on self-organization: "Life is an anticipated collective asset of catalytic polymers for self-organization." [5].

The definitions by Joyce and Kauffman focus on chemical systems, which consequently exclude technical forms of self-organization. Kauffman's definition, however, would allow the thought experiment involving molecular soup in a larger vessel to be considered life. The robot community, which ultimately could be created by humans, would from a biological point of view almost approach what we consider to be life.

From the point of view of astrobiology, the search for a definition is important, because, in the search for life in space, the question could arise as to what signs of life we can accept as such (see more in Chap. 9).

1.3 Who Was LUCA?

On the basis of biochemical data, we have good reason to assume that all living beings on earth descend from just one ancestor. It must have been a cell that managed to grow and divide for the first time, actually leading to surviving daughter cells. The descendants needed to survive until they themselves divided again—a process which continues to this day. This first cell is called LUCA (Last Universal Common Ancestor), the last common ancestor of all living plants, fungi, and animals, including humans. For LUCA to form, a continuous production of molecules must have taken place long before that, which provided the necessary basic building blocks for the experiment we call life. These include organic bases, such as adenine or guanine, and amino acids or the lipids required for building cell membranes.

But building blocks alone are not enough. Spaces for reaction were also required where attempts to assemble more complex connections could take place. Small caverns or pores were sufficient in which the molecules could accumulate. Their concentration must have been at least high enough for them to meet and react with one another sufficiently frequently. A very large number of tiny labs were required, all linked to one another against a background of changing conditions, material replenishment, and the disposal of unusable components. That said, however, high-molecular concentrations also pose a new problem: the variation and the number of different molecules are so large that special selection processes are required to crystallize functional connections for life. The biological cell LUCA must have formed under such conditions as the most successful system that has ever been created on earth. From then on, planet earth entered into a unique development.

The composition of the atmosphere changed significantly as a result of photosynthetic bacteria and plants. While the level of carbon dioxide (CO_2) reduced, the oxygen content rose continuously. Organic acids and later on plant roots and the activities of animal contributed to increased weathering. On the one hand, this resulted in increased erosion, but, on the other hand, with the onset of soil development and the formation of plant cover, a delay in erosion processes. This changed the water balance in the rivers, and also the type of sediments and their transport. Organogenic sediments such as coal and reef limestone were formed, which again had a direct influence on the composition of the atmosphere through the carbon dioxide balance.

And finally, human beings appeared on the stage, who brought about changes over a short period of time, the scope of which can only be compared to the impact of a large meteorite. Ultimately, everything we see today on the hard surface of the earth is the result of the successful propagation of LUCA. Without LUCA, even mountains would look different today, free of biogenic limestones and oxidized iron minerals, manifesting other forms of erosion, and free of lichen and bacteria films. One small exception may exist that is still evident to us, however. These are the very young volcanic structures that protrude from the earth void of vegetation. But here, too, LUCA has had its fingers in many pies. The frequently red surfaces of the lavas, which contain ferrous minerals that have been oxidized by atmospheric oxygen, can be seen from afar. They are evidence of a change in the atmosphere that began more than 2.4 billion years ago, when the mass production of oxygen caused by cyanobacteria led to a constantly increasing concentration in the atmosphere.

1.4   The Beginnings

"How did life actually come about?" This question was posed to the small group of scientists who had gathered for the first time in the university cafeteria in Essen. Everyone looked at each other and shrugged. The opinion of everyone present was unanimous: "It's much too long ago to find out; we can only speculate; there's nothing really tangible; the general conditions are hidden in the fog of the past". "But's that the reason why we're meeting now, so we can talk about it in the first place," interjected one of the colleagues present. "I have a slight suspicion", I offered tentatively and began to sketch the model of a tectonic fracture zone on a napkin with a pencil…

Haven't we all already asked ourselves the question of how life originally came about? Certainly, those of us who see consider the laws of science to be the basis for our existence. So far the answers are vague. And it's not just about the beginning of life. It is just as important to clarify the phase in which the planet was born from which the preconditions were created that could give rise to life. Looking back even further, the birth of the solar system is also of fundamental importance. Its development is associated with decisive influences that still determine fundamental processes on earth. The beginnings go so far back that we know only too little about the earth in its infancy or about outside influences and processes inside the earth and on the surface. Numerous hypotheses exist about life, none of which is generally accepted. The group quickly came to the conclusion that the question of the origin of life is one of the most complex questions in science. One that is unsolvable! So why don't we just let the matter rest?

No, we certainly won't—thus the unanimous opinion of all those present, who continued to meet me on a regular basis after the first cafeteria meeting to discuss this new idea about the origin of life—with the attraction of researching something completely unknown, with its infinite number of question marks, as the motivating factor. This unified the group without any obligation to deliver results by any specific deadline, satisfying our curiosity alone, and perhaps simply identifying a small segment in the process and taking a first step toward a possible solution—that was certainly worth it.

From a scientific point of view, the search for an introduction to the origin of life immediately made it clear that nothing else we know matches the complexity of life. Life encompasses our entire earthly world view. In its complexity, its development has gone so far that simple attempts to explain individual processes seem almost impossible. Time was needed for this development up until the present day, an expanse of time that infinitely exceeds every human horizon in terms of experience, perhaps 3.5 billion, maybe 3.8 billion years, or more. This vast period of time, which was apparently necessary for an abstract thinking being to develop, makes an introduction to understanding the infancy of the earth up to the present day so difficult. The initial phase is immersed in a thick fog that seems impenetrable. Too many unknowns exist up until the present that make the environment in which life was created so complicated. What exactly do we know about the conditions on earth in its infancy? What was the primordial atmosphere, or the waters of the primordial oceans composed of? What proportions of organic molecules came from space transported by meteorites or comets; how much land surface existed and up until when? What influence did the moon have after it was formed?

In addition to planetary and geological unknowns, those relating to physical chemistry and biochemistry all exist. Which processes contributed to such a high-molecular concentration that allowed reactions from the simplest chemical building blocks to complex molecules to take place over long periods of time? How were these building blocks linked on earth—in an environment containing water, which is defined as a general precondition for life but to an extreme extent hampers reactions? They only take place today with the help of enzymes in the aqueous environment of the cell or in the lab in an organic solvent. For the earth in its infancy, only organic solvents like alcohols or ethers in very low concentrations are conceivable. Above all, we have to answer the fundamental question of how the chemical storage of the information came about that is contained in every DNA and which carries the entire development of the biochemical processes over a period of more than 3.8 billion years. One data chip in every cell: can't we just read it out and identify the beginnings like that?

The answer is a resounding "no". It is not without reason that many statements on this topic by past researchers echo the resigned statement that it will probably be impossible to fully determine, and let alone explain, the processes that lead to life.

1.4.1   But How Did It All Begin?

Every research project is underscored by a history, one longer and the other shorter. The history of the research into our own origins began at the end of the 1980s in Westerwald while processing the 20 million year old volcanoes in this region in the Rhenish Massif in Germany. In the course of investigations, structures became apparent that could only be explainable through special supra-regional tectonic processes and fracture structures in the crust, but this could not be clarified. It was only later on, after the turn of the millennium, that the opportunity came about to investigate the formation of fracture structures further in the neighboring region of the Eifel. The mapping of tectonic fault zones, which provide a vertical gas-permeable connection to the earth's mantle, delivered a surprise. Every time fracture zones and the escape of gas—mainly of carbon dioxide—were identified, the presence of hill-building forest ants was identified locally at the same time. The correlations were so obvious that, based on personal experience, predictions about forest ant sites could be made based on geological knowledge alone, an absurdity in biology! The observations gave birth to a new field of research that sparked many discussions. Initially, there was rejection from both sides representing geology and ant research, which, after a long period of intransigence, eventually led to a successful collaboration with entomologists. The search for the causes as to why the representatives of the forest ant genus *Formica* settle on gas-permeable fault zones led to considerations in all directions. Is it the moisture, heat, biofilms in the fissures, or substances that rise to the surface besides the gas and possibly feed the ants or bacteria in the biofilms? Does the CO_2 present help to prevent the fungus on eggs and larvae or keep enemies and parasites under control at high gas concentrations which they are able to tolerate? Or does the carbon monoxide that rises in small concentrations along with the carbon dioxide have any function at all? How is it removed from the hemolymph, the circulating fluid, or "blood" of insects? Do forest ants use this to form formic acid—their chemical weapon—with a molecule of water?

The more extensive realm of the fault zones came more and more into focus with all these considerations. Weren't all the raw materials required for forming organic molecules present here? The pressure and temperature conditions can also be found to exist there that are defined in technical-chemical processes, such as those found in the Fischer–Tropsch synthesis to form hydrocarbon compounds (see Sect. 3.1). This synthesis can, for example, be used to produce synthetic gasoline. Enough metal compounds also exist that can serve as a catalyst. Could all these things possibly be the key to the ants' preference for building their nests on these faults?

And then suddenly it clicked! These represented the ideal conditions for the early stages of life: a protected space, available for millions of years, with all the raw materials required and an infinite number of small reaction spaces, in which different pressure and temperature conditions and pH values existed and still exist today.

The idea for developing a model for the origin of life was born. From that point onward, the evening meetings between Oliver Locker-Grütjen and his scientist colleagues came into play, and these continued to take place over a period of more than 10 years. Representatives from the fields chemistry, biology, physics, physical chemistry, bioinformatics, microbiology, and my discipline, geology, met. It was the relaxed atmosphere with food (usually cooked by Hans-Curt Flemming) and wine and beer that ultimately helped to let all those gathered discuss the question that concerned everyone. And then following the disillusionment, none of us really knew anything about the specific question being asked. Certainly, the common knowledge around the beginnings of this discussion was known, the old ideas and the initial investigations. Even the newer developments, which had been repeatedly published in the media, were familiar to us. But the actual core questions concerning cell formation, information storage, and enzyme formation, like for all our colleagues worldwide, remained completely in the dark. As a result, we began to research, lectured each other about newly learned subject matters, invited colleagues who already made a name for themselves in this field of research to colloquia, and even planned initial experiments designed to have something to do with a tectonic fault environment in the continental crust.

The suspicion arose more and more, however, that the field resembled an impenetrable fog with no beginning or end. The progress our considerations were making slowed and seemed to be approaching a limit. In a side note, I attempted to ascertain the meaning of CO_2 in the development of molecules like amino acids or organic bases. While doing so, I remembered a paragraph from the book *Chemical Evolution* by Horst Rauchfuss [6]. At his advanced age, we even dared to invite him to a lecture in Essen, which he gladly accepted. The paragraph in his book addressed reactions in CO_2 excesses which are meant to be beneficial for the formation of organic molecules. Since no one could remember the passage, I researched it on the Internet the next day. Even the first links returned by the search query contained information in their sublines that hit me like a bolt of lightning and changed everything. When the temperature rises above 31 °C (304,15 K), CO_2 is present in the supercritical phase down below a crust depth of about 740 m (73,75 bar).

This fact, which was taken for granted, had simply disappeared in the enormous fog we were poking around in. I just reported my new discovery to Christian Mayer. As a physical chemist, he immediately knew that this gave us a tool that opened up a whole new world, a solvent which we could use to carry out reactions that were impossible on the earth's surface.

Further coincidences also occurred that led to the point where we are today. One of them was the introduction of a new software system for the university's administration. This led to the financial situation for all the faculties seemingly disappearing from sight into a kind of primordial fog for more than 2 years. During this period, considerations concerning the purchase of a high-pressure system matured. Since our traditional sources of finance persistently refused to support our research, I purchased the system we required from supposed personal financial reserves. It represented the most important action taken in all my research. Each experiment undertaken brought results that represented something completely new and were a big step toward understanding the processes in place when life began. Once the financial fog had cleared, the precise six-figure sum that the system had cost had accumulated as debt in my household account. Without this fog, I would never have dared to incur such an enormous debt, and the experiments would never have been undertaken.

Another circumstance was very fortuitous as well. Whereas time constraints forced the group to be reduced to just a few participants, a new colleague, Oliver J. Schmitz, joined from the field of analytical chemistry. His new, state-of-the-art lab and his immediate willingness to work on the project offered us the first opportunity we had to analyze the low-molecular concentrations from our high-pressure experiments carried out by Maria Davila-Garvin. The series of experiments undertaken by Christian Mayer on cyclic vesicle formation led to a special analytical challenge, which the analyist Amela Bronja ultimately took on and with success.

A multitude of lucky coincidences of a small and large nature also occurred that ultimately led to contacts and progress being made, without which many aspects would have become bogged down in the introductory phase. This includes the connection in the field of geosciences to Heidelberg to the research groups under Heinfried Schöler and Frank Keppler, who had a significant involvement in the investigation of the fluid inclusions in hydrothermal quartz (along with their employees, Ines Mulder, Tobias Sattler and Markus Greule, and Mark Schumann from my research group), and to Jonathan Williams from the Max Planck Institute for Chemistry in Mainz, who cultivated a whole network of important contacts in the USA. Gerald Dyker from Bochum, as an organic chemist, also provided several suggestions for further experiments. Last but not least, I need to mention the cafeteria principle, which made direct exchange and "further education" possible in a subject that was relatively foreign to me. The physical chemist, Christian Mayer, the biochemist, Peter Bayer, the bioinformatician, Daniel Hoffmann, and the analyst, Oliver J. Schmitz, were all targets for my barrage of questions and sometimes excessive ideas. But it brought results: the cafeteria meetings followed by coffee proved to be the most effective form of scientific exchange.

If, like us in the Essen Group, you are addressing the topic of the "origin of life" for the first time, the first thing that arises are fundamental questions concerning the state of science. What do we know so far, what considerations did past researchers have, and what experiments were undertaken that provide hint at the beginnings from which life may have developed?

A number of natural scientists have dealt intensively with the question of the origin of life over the past century. It all started with a cautious approach to the complex topic, which could only be treated theoretically at the beginning. Experiments did not begin until the middle of the last century, which gave the first indications of the possibility of a biochemically based start to life. As expected, the success of early research could only be very limited. Little was known about the general conditions, such as the planetary and geological development of the earth or the influence of astronomical quantities. The compilation of older works listed in the later chapters in this book provides insights into how every time new discoveries were made new ideas were developed. Chapters 2–4 summarize the current state of knowledge in brief, which focuses on the planetary and physicochemical preconditions. This alone results in a multitude of important fringe conditions for the development of life as we know it. The essentials of the models made known to the general public in recent decades are presented here. These take the form of explanatory approaches that offer a broader basis and contain more than just one section in an interesting series of reactions. The consideration takes place in the context of more recent planetary knowledge. We need to take previous researchers into account here who were forced to take the current state of knowledge as their starting point back then, which, of course, placed limits on the statements made in their models.

In recent years, research on the origin of life has received an increased impetus around the world. New findings from the analysis of meteorites, the ocean realm, and continental crust have led to new model ideas, which make it possible to perform laboratory experiments for the first time under robust general conditions. Furthermore, documents from the early days of the earth have been discovered that demonstrate the beginnings of organic chemistry. All the data now available means that we can see a first blurred outline in the fog of the past, which makes one thing clear: the riddle about the origin of life can be solved. However, it is already becoming apparent that it took a physicochemical process over an extremely long period of time to develop the cell, which is considered the last common ancestor of all life on earth.

In the evening discussions undertaken by our "Origin" group, one aspect existed that resonated latently in the background: if we succeed in developing a notion about the early stages of life, does that mean we can also transfer this model to other planets in the universe? Or even crazier than that: can we conclude from this that there are other planets in space with some form of life, perhaps a higher form? It soon became clear: it was not out of the question. But the chance of an intelligent developing will quite so often not be found elsewhere.

References

1. Bojowald M (2009) Der Ur-Sprung des Alls. Spektrum der Wissenschaft 5:26–32
2. Bojowald M (2009) Zurück vor den Urknall. Die ganze Geschichte des Universums. S. Fischer, Frankfurt am Main

3. Feinberg G, Shapiro R (1980) Life beyond earth: the intelligent earthling's guide to life in the universe. William Morrow, New York
4. Joyce GF (1995) The RNA world: life before DNA and protein. In: Zuckerman B, Hart MH (eds) Extraterrestrials. Where are they? Cambridge University Press, Cambridge, pp 139–151
5. Kauffman SA (1996) At home in the universe: the search for laws of self-organization and complexity. Penguin Books, London
6. Rauchfuss H (2008) Chemical evolution and the origin of life. Springer, Heidelberg

2

Global Requirements

Abstract Basic preconditions are required for the emergence of life, including rock planets in a solar system, which are in the habitable zone at the correct distance to the sun. Furthermore, the solar system needs to be located at the edge of a galaxy. Organochemical compounds on the earth's surface need a magnetic field to protect against the solar wind to survive. Over and above this, a rotation of the planet and an accompanying moon that stabilizes this rotation to provide a harmonious heat balance are favorable. The moon was formed a few million years after the earth was formed by colliding with a minor planet. The subsequent cooling of the earth led to the formation of a magnetic field and a first atmosphere. Water was created during the late phase of continuous bombardment by meteorites from the outer area of the planetary system.

2.1 Initial Requirement: The Planets and a Sun with System

Where should we begin? The framework factors that do not directly relate to life seemed to be better known than the start of organic chemistry. So, we went back to the time before the beginning of life, to everything that had something to do with the planetary conditions. We could certainly have included the initial phase of the universe immediately, but that seemed a bit far-fetched in relation to the actual development of life on earth. Besides, it is also an aspect that is even more obscure than the origin of life itself. Only

© Springer Nature Switzerland AG 2020
U. C. Schreiber, C. Mayer, *The First Cell*, https://doi.org/10.1007/978-3-030-45381-7_2

hypotheses for the start will ever exist, since experiments able to provide evidence are impossible. The actual processes involved in the development of life can be seen independently of this as physicochemical and biochemical processes. What has to be assumed is the existence of matter at a certain point in time, which made the formation of galaxies and solar systems possible. When we look further, our solar system inevitably comes into focus. So far, planet earth is the only example we have available for the development of a biological cell made primarily of inorganic matter. We conclude from our knowledge that a rock planet, accompanied by a moon, if possible, within a solar system is required as a precondition for life. At the same time, the position of this system within the galaxy is important. In the center of each galaxy, supernova explosions or neutron star collisions take place over the entire lifespan of the respective galaxy. These create high-energy gamma-ray bursts that pose a danger to all life. The danger decreases toward the outside in the direction of the spiral arms, where our solar system has also developed.

The sun itself, as the center of a planetary system, needs to be of a certain size. This has a direct effect on their lifespan, the luminosity, and the strength of the gravitational forces on the planets. A planet which should bring chances for the development of complex organic molecules needs to be located in the solar system's habitable zone and demonstrate certain characteristics. Astronomers define a habitable zone as a relatively narrow area around the sun, the distance from which is so large that water can exist in liquid form on the planets that exist. Water in liquid form is an essential requirement for life as we know it. Too great a distance from the central star results in temperatures below freezing. With greater proximity, an atmosphere that forms including water vapor will be destroyed by high temperatures and/or radiation pressure (solar wind). One keyword phrase describes the situation of the earth very clearly: it is located exactly at the triple point of water. That means that ice, as well as water and water vapor, i.e., solid, liquid, and gaseous water, exists. But this is only true to a limited extent, however. The earth is actually so far away from the sun that it should have been orbiting it as an ice planet from the very start. But following a transition period, a lucky coincidence came into play: a gas envelope formed, which, with its heat-reflective properties, prevented surface temperatures on the earth from permanently remaining below zero.

A rotation of the planet and an accompanying moon, which stabilizes this rotation, are also favorable to the heat balance. Rotation compensates for the extreme climatic conditions that arise when the side facing and the side facing away from the sun come to a standstill. Coupling the rotation with a moon prevents extreme fluctuations in the inclination of the axis of rotation, which would have a major impact on the climate. Raw materials are also required that lead to carbon chemistry in addition to the existence of water. The possibilities for developing alternative life forms with components other than carbon, hydrogen, and oxygen, such as on a silicon basis, for example, are considered to be extremely low. However, we only have experience of a carbon biology, which is based on our limited, earth-related view. What is more, the early development of life requires a magnetic field that fends off the particle stream constantly emitted by the sun. I will go into the meaning of the last point again in more depth in a separate observation.

2.2 Second Requirement: The Earth—A Collection of Materials for a Beginning

Life on earth is a sure and visible sign that all the general conditions required for its creation were present at the right time. Once the central star in our solar system had been formed, sufficient dust material was available in its environment to allow larger objects to be created as a result of continued collisions. The particles melted into smaller and larger meteorites, asteroids, and planetesimals and kilometer-sized asteroids that eventually became larger and larger objects. At the end of it all, the rocky planets existed that remain to this day. The formation of the first planetesimals began 4567 to 4568 billion years ago [1]; the accumulation into planets controlled by gravitational forces, which is called accretion in astronomy, took place during the following 30 to 100 million years. The time when the moon formed is of importance. According to the latest calculations, it was formed more than 4.5 billion years ago, during a 20 million year time window after the formation of the earth. An intensive discussion exists around the conditions for its formation, to which alternative models have given new impetus. Most support is given to a model that sees the infant earth colliding with a planet the size of Mars, called "Theia," as the cause [2]. Here, debris from both planets are said to have been hurled into space, which then united to form the moon. A variant of this model brings more energy into play that could solve previously unresolved problems. On the one hand, the isotope ratios of the rocks of the moon are almost identical to those on earth, but on the other hand, sodium and potassium appear to have been lost in the formation of the moon. Geophysicist Simon Lock from Harvard University has postulated a highly regarded hypothesis. According to his theories, the collision between the earth and Theia was so violent that the entire material or at least a large part of both planets evaporated and then condensed again to form solid matter. This resulted in the formation of the earth and the moon in tandem according to his calculations. The hypothesis could help explain the same isotope distribution of different elements on earth and the moon and the loss of potassium and sodium in moon rock [3, 4].

A statistical analysis of the impact craters on the moon shows that a phase between 4.1 and 3.8 billion years must have existed during which the planets and the moon were struck by an increasing number of asteroids (heavy bombardment) [5]. Every time the fragments collided, the kinetic energy was mainly converted into heat. This means that growth into a planet like earth was only possible in connection with being strongly heated up. The temperature reached such high values that parts of the rock masses became molten but not the entire planet. According to older calculations, it took around a billion years until the rise in temperature caused by the decay of radioactive isotopes was so high that under the prevailing pressure conditions, iron began to melt at depths of up to 1000 km. Estimates of the development in temperature show that this was the only way of separating the material into an iron-rich core and the rocky shells of mantle and crust. This situation still describes the distribution of materials inside the earth today. However, when the ideas concerning the formation of the moon became more concrete, these calculations were questioned again. If a collision with another planet did take place after the earth was formed, this was a clear turning point in the history of the young planet cooling. Whether an abrupt rise in temperature, which is to be expected after such a collision, led to partial or complete melting of the earth is currently a matter of debate (see introductory paragraph to Sect. 2.2). The formation of the earth's core is key to the origin of life. Its creation from a mixture of liquid iron and nickel laid the foundation for the development of the earth's magnetic field [6].

With the first cooling, and even before the moon was formed, an initial crust must have formed from a high-temperature basalt that was particularly rich in magnesium and low in silicon and aluminum. This basalt type is called komatiite. If the collision model is confirmed, the first crust was destroyed by the collision with the Mars-like planet. With renewed cooling, a crust formed that consisted of the high-temperature basalt komatiite. Only a few relics of komatiite still exist on the earth's surface today. Nothing is known about an atmosphere before and immediately after the moon was formed. Enough sources of gaseous substances existed, however, some of which came from the interior of the earth. They rose to the surface through cracks and volcanic eruptions. Others came with the meteorites and comets that followed from a belt that was too far from the sun to allow the volatile components to evaporate. At that time, the earth's gravitational pull was actually sufficiently large enough to stop water vapor, carbon dioxide, and nitrogen escaping into space. Nevertheless, an atmosphere that had existed for longer has to be questioned, since presumably a sufficiently strong magnetic field capable of offering protection from the solar wind had not developed inside the earth.

The planetary magnetic field is the reason why the stream of particles continuously released from the sun (solar wind) fails to reach the earth's surface. Today, only after particularly strong eruptions on the surface of the sun do charged particles in the polar regions come into contact with the atmosphere where they excite particles that cause the northern lights (aurora borealis). A weaker magnetic field, as was the case in the early phase of the earth's development, moved the magnetopause closer to the earth. The magnetopause is the boundary where the dynamic pressure of the solar wind equals the dynamic pressure of the magnetic field, i.e., a barrier that does not let the plasma flow through but deflects it to the side. In the case of a magnetopause close to the earth, strong eruptions on the sun's surface can, however, cause the plasma flow to break through the boundary and reach the earth's surface. It has to be assumed that parts of the young atmosphere were eroded again and again by such events. During the initial phase, the sun emitted a solar wind that was a hundred times stronger than today. At the beginning, this particle stream struck the unprotected atmosphere, which was still in formation and had no magnetic field and could not withstand it. It was swept away, like that of Mars, which no longer has a meaningful atmosphere today. The atmosphere on Mars has less than one-hundredth of the atmospheric pressure of the earth's atmosphere.

From the beginning, simple organic molecules were able to develop in low concentrations on land masses above the surface of the water. This took place through surface water coming into contact with hot lavas and gases or through reactions that took place owing to the high-energy radiation from the sun. Furthermore, a small amount of organic molecules from space existed. All organic molecules exposed to the solar wind were destroyed after a short while. The very strong ultraviolet radiation had the same effect, and at the beginning of solar activity, the proportion of radiation was much higher than today. This was exacerbated by the fact that a protective atmosphere with an ozone shield still did not exist, which could only be formed in a sufficient thickness in the upper layers of the gas envelope after biological oxygen production began more than 2.4 billion years ago (Fig. 2.1: O_2 in the atmosphere).

Some of the oldest rocks on the surface of the earth can be found in Northwestern Australia. In addition to quartz, sedimentary rocks often contain extremely stable zircons, which in turn can have magnetic inclusions.

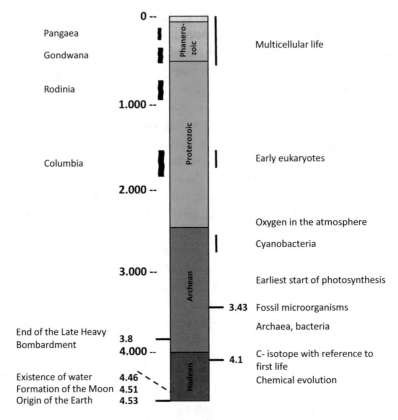

Fig. 2.1 The earth's epochs and significant events and the appearance of life forms in the timeline. Upper period (yellow) = Cenozoic. Black wavy bars correspond to times for the major continents (Columbia, Rodinia, Gondwana, and Pangea)

They can be used to identify evidence of a magnetic field present at the time of the inclusion. Studies of such mineral inclusions from Jack Hills in Western Australia indicate that a magnetic field may have existed 4 billion years ago, but with a strength of only about 12% of today's magnetic field [7]. However, the values determined are associated with greater uncertainties owing to influences related to temperature during subsequent overprinting of the rocks with high temperatures and pressures (rock metamorphosis). From minerals in younger rocks from 3.4 billion years ago, field strengths could be measured that correspond to at least 50% of the earth's magnetic today [8]. Despite greater uncertainties in the data obtained, this shows that the buildup of the magnetic field required a longer period of time. Due to its initial low strength, considerable problems existed with the development of organic molecules on the surface of the earth, along with their preservation and the formation of reaction chains, which could lead to more complex molecules.

> **What's So Special About the Atmosphere on Mars?**
>
> The missing atmosphere on Mars is said to be related to another development in the planet's core. This was also liquid at the beginning, but quickly became solid. Accordingly, a magnetic field existed during the liquid phase that shielded the solar wind. At that time running water and an atmosphere existed. After the molten iron crystallized in Mars' core, its magnetic field collapsed. This was followed by the attack by the solar wind, which swept away the atmosphere and the water, which evaporated [9, 10]. Theories exist that the infant planet Mars also provided the right conditions for life to originate, but that the conditions described prevented the development of lower life forms.

2.3 Third Requirement: Water

"How did water find its way to earth? This question came up relatively early in the discussion about the origin of life: astronomers cheer when they identify water on other planets or moons, even when it is only the slightest trace. It seems to have been a very special stroke of luck for us to have got so much at once for even two-thirds of our planet to be covered with water. Water is the source of all life. But where did it come from? The processes that led to the development of the planets are associated with such high temperatures that the formation of water during the initial phase was impossible. This means it must have come from outside, like everything else—only a little later when the temperatures were no longer so extraordinarily high. The comets, with their high proportions of H_2O and CO_2 ice, have been discussed as potential candidates. However, an analysis of their isotope composition brought disillusionment. Almost all water molecules investigated from the comets differ significantly from the isotope composition of water on earth. Only a small amount can be derived from them. Besides the consideration that maybe the right comets with the corresponding isotope ratio have not yet been found, other water bearers could also be identified.

Modeling the orbital movements of all planets since their formation revealed that deviations in the orbits of the large gas planets Jupiter and Saturn in the early phase of the solar system must have caused turbulence in the inner and outer asteroid belts. While the asteroids close to the sun contain only a small amount of water, planetesimals with up to 10% water occur in the outer belt. The orbital disturbances of the asteroids within the belts created by the gas planets caused some of them to be deflected onto a collision course with earth. The calculations showed that a contribution of 1–2% of the earth's mass from this outer belt is sufficient to explain the amount of water on earth today [11]. According to these ideas, the main mass of water only reached the earth in the late phase of the bombardment to which all planets were exposed. Large quantities of water ended up in the earth's mantle, from which small portions are still released into the atmosphere today during volcanic eruptions (Fig. 2.2).

2.4   Fourth Requirement: A Permanent Atmosphere

We therefore have an idea of how the planet was formed and how the rocky sphere, the lithosphere, was formed. Plausible explanations as to how the water that formed the hydrosphere came about also exist. Now only the third factor is missing, the atmosphere. All three form an intersection that has something to do with the development of life. The formation of a constant atmosphere on a planet is linked to three essential requirements. First, substances need to be present that are gaseous under the pressure and temperature conditions on the planetary surface. Secondly, the mass of the planet must be large enough for gravity to stop the gases escaping into space, and finally, it must be possible for a magnetic field to develop that stops the particle beam from the central star (solar wind) reaching the surface. Today, the composition of the earth's atmosphere is significantly influenced by organisms

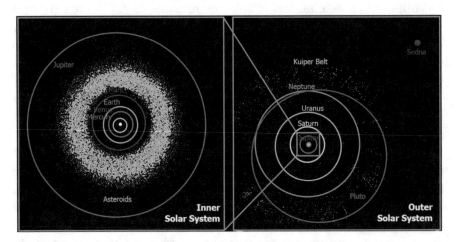

Fig. 2.2 Inner and outer solar system with planets and the asteroid and Kuiper belt (©Courtesy NASA/JPL-Caltech [12])

that drive photosynthesis. The appearance of oxygen in the atmosphere is documented for the time from 2.4 billion years ago before today. This provides us with a unique feature that is unknown on any other rocky planet. The actual phase when the earth's atmosphere began can no longer be recognized today, because it was destroyed by the solar wind. However, the suppliers of the gases can be identified in part, which, as in the case of water, also originate from the asteroid belts. This means they belong to the later phase of accretion, which ended more than 3.8 billion years ago. That said, however, the proto-earth also harbored enough raw materials from which gases, such as hydrogen, ammonia, carbon monoxide, and carbon dioxide, could be formed. They were partly dissolved in the molten rock in the earth's interior and slowly released into the atmosphere by volcanic eruptions and through discharges at fracture zones in the infant crust. One can only speculate about the proportions of the individual gases in the first stable atmosphere. Water vapor certainly existed that made up a large part of the gas. However, the order of magnitude and the ratios of carbon dioxide (CO_2) to carbon monoxide (CO) and nitrogen (N_2) to ammonia (NH_3) are unclear. Besides methane, hydrogen, and sulfur compounds (H_2S, SO_2, and sulfuric acid [H_2SO_4] as aerosol), noble gases and traces of other gases also existed. Depending on what predominated between CO_2/CO and N_2/NH_3, the atmosphere was either weakly or rather strongly reducing. Strongly reducing in this context means that chemical reactions are facilitated by the absorption of electrons by reaction partners that under weakly reducing conditions would not release these electrons. This can be crucial to intermediate steps in the formation of complex organic molecules. Lammer et al. provide an overview of the state of the debate on this [10].

The concentration of carbon dioxide (CO_2) in the early atmosphere is not known for sure. Estimates exist which lie far apart and extend beyond the content of the present proportion of oxygen. Common to all considerations is that at the beginning, the value is assumed to be far above that of today's concentration. An approximate size for this is the estimate for the volume of the sedimentary limestone deposits on earth.

2.4.1 What Does Limestone Have to Do with the Atmosphere?

A lot! Almost from the beginning of the development of the atmosphere, a very effective chemical chain reaction started, which resulted in a reduction in the CO_2 concentration in the gas mixture: the reaction by carbon dioxide with calcium ions from the rock in the earth's crust. The earth's crust contains 4% calcium. It represents a key element in the composition of igneous rocks and is integrated into the crystal lattice of minerals together with oxygen, silicon, aluminum, and other elements. These include feldspars or pyroxenes. The bright parts on the moon are made up, for example, of a powdered rock (anorthosite), which to a large extent consists of calcium-rich feldspar (anorthite). The same minerals also occur on earth. The weathering of rocks on the earth's surface and contact with calcium silicates by CO_2 dissolved in seawater produces calcite ($CaCO_3$) from CO_2 and calcium (Ca) through a series of intermediate steps. It was only in the course of the earth's development that the evaporation of seawater in shallow sea basins led to lime precipitating, thus forming a chemical sediment. This is a process that we experience in regions where hard drinking water comes out of the tap. We can safely deduce from the type of sedimentary limestone formation that at the beginning no sedimentary limestone existed on earth. Today, large sections of alpine rock, the Alb, and the chalk on Rügen and in Denmark and England and many regions around the world mainly consist of limestone with the mineral calcite. If we take all the corresponding limestones on earth to calculate the CO_2 integrated into the crystal lattice, we get an idea of how much CO_2 mineral formation has removed from the atmosphere. However, we also need to take into account here that CO_2 degasification of the earth continues to this day, and the volume of the quantity leaked out over 4 billion years also needs to be taken into account accordingly. Today, the annual contribution being added by the earth is in the order of 1% of man-made CO_2 emissions [13]. At the same time, other mineral formations exist as well, like iron, which have bound large amounts of CO_2 from the atmosphere.

2.4.2   Please Note

We need a mind map at this point, a kind of white board like you would see in any crime film to map ideas thoughts, ideas, and connotations. As you know, the white board is used to pin up photos of suspects and group clues so that a map can be built using cross-connections that helps to outline the course of the crime. We place the earth on our white board, which, if we take its distance from the sun, should actually be an ice planet. We also place the still weak sun, which is plagued by high radiation (ultraviolet radiation and solar wind; see Sect. 5.7), with a place for the moon formation process and also one for the magnetic field, which initially develops over time. A little further down the board we place the atmosphere, which we connect to the earth's magnetic field and the sun. At the same time, we link it to a field for the greenhouse effect that expands and enlarges the solar system's habitable zone outward. This then forces the earth to depart from its position as an ice planet. And in the middle we place a big question mark. How could a heat-protecting atmosphere form under the influence of the solar wind, which kept the planet ice free after cooling, if the earth's magnetic field only formed gradually? No clear statements are available on this yet. In any case, the models that assume that life developed on the face of the earth also need to answer the question of it being completely iced over during the first hundreds of millions of years of its life.

2.5   What Happened Next?

This is exactly the question we asked ourselves after researching the initial conditions. The answer to this question was of no small importance, because we wanted to discuss the different models for the origin of life, which require the earth to have a solid surface from a certain point in time. The earth's crust provides the space required for reactions, an interface to the hydrosphere and atmosphere, is connected to the earth's mantle over fracture zones, and provides the requisite resources. In later chapters of this book where I present fresh considerations of the development of the first cell, the existence of a continental crust on infant earth takes a central position.

At first, as it was made clear by ideas concerning the magmatic processes on infant earth, only a basaltic crust existed that formed with the cooling after the proto-earth collided with a minor planet. It was very thin (perhaps just a few hundred to a thousand meters thick) underneath a highly active mantle featuring strong convections. The mantle-crust system must have started to separate the substances for forming a continental crust early on. They formed the core zones for the first continents which grew continuously over time. A still active example is Iceland, which owes its existence to its special location on two plate boundaries and a superposition of two magmatic processes. Before I provide you with a more detailed description of the conditions prevalent there, I would like to go into the basics of all-dominant plate tectonics. By doing so, I can best convey an understanding of Iceland's special features.

One topic which has long been in discussion among geoscientists has to do with the point in time when plate tectonics began, which are still active today. This played a crucial role in the course of the earth's development and the development of higher forms of life. First and foremost, it stands for the formation of large mountains, the weathering of which led to the CO_2 binding sustainably. A reduction in CO_2 had an immediate impact on the climate. Over and above this, sedimentation spaces developed, which absorbed the sediments from both the mountains that eroded and shelf areas where the most important steps in the development of higher life forms took place in later times. Plate tectonics took on a particular importance after the continents settled, with habitats being separated and merging elsewhere. Added to this are influences on the marine area through chemical exchange processes in the oceanic crust, the ingress of metallic solutions in the oceanic ridge, and much more. A brief description of the essential factors would therefore be helpful to further consideration.

The engine driving recent plate tectonics is made up of a combination of several factors. A moving plate has various boundaries. On one side, it is formed by the growth of basaltic rock masses. The basis for this growth is formed by magma, which rises, with some of it getting stuck in the production channels and crystallizing and some flowing out onto the ocean floor in the form of lava. The lava forms a special type of basalt called pillow basalt. This plate edge is located on the oceanic ridges, onto which the plates that drift apart laterally are added in equal portions. The magma for this comes from the mantle where slow convection currents generate an equalization in volume. The oceanic plate in the vertical always consists of two parts today and features basaltic chemistry, an upper part—the oceanic crust with feeder dikes and the pillow basalts (on average 10 km thick)—and a lower part, which is approx. 90 km thick that forms part of the upper mantle. The latter consists of a solid rock subject to the pressures and temperatures that prevail at this depth. Only the mantle that is even deeper is so hot that it has a low melt content and is therefore able to flow slowly. The temperature differences between the lower and upper mantle drive the slow circulation that mixes the mantle (Fig. 2.3). It can be compared to the water that circulates in a pot standing on a hot stove. The thickness of the plate on the oceanic ridge is still very narrow, perhaps only a few kilometers thick, consisting of the crust and a firm upper mantle. Only with a greater distance from the oceanic ridge does the plate cool down more and more, which causes the lower part of the plate to thicken. In this way, the most distant parts of the plate achieve the greatest age and thickness of more than 100 km. The magmatic processes at the plate's linear point of origin cause the entire oceanic ridge to rise above the average level of the oceanic crust. This results in a lateral slide caused by gravity, which affects the entire plate. The opposite edge of the plate is determined by it being immersed in the mantle, along a zone called the subduction zone. Deep-sea trenches form along the subduction zones and volcanic belts form above the submerging plate, which are located at a characteristic distance

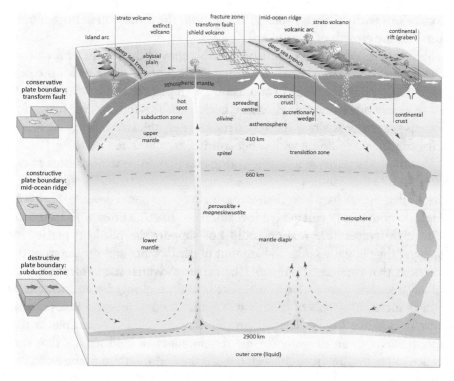

Fig. 2.3 Schematic representation of the plate tectonic processes on the earth's surface and in the earth's mantle. (© Springer-Verlag GmbH [14])

from the subduction zone (Fig. 2.3). The volcanic belt is already part of the neighboring plate, where fault zones occur owing to powerful tectonic tensions. Strong earthquakes take place at these fault zones. The immersed plate retains the greatest thickness owing to the long amount of cooling time on this side. At the same time, its density is higher than the hot underlying mantle material. As a result, it sinks into the plastically deformable mantle, like a paving slab that slides into a depression filled with morass. A pull is exerted on the subsequent mass of the plate, which supports it sliding from the relatively higher position at its place of origin. Convection currents in the upper, plastically deformable mantle also act as an influence here if they run in the same direction as the one in which the entire plate is moving. Since a part of most plates is formed out of oceanic crust and a part out of continental crust (e.g., the African plate, consisting of the main part of the African continent and west to the mid-Atlantic ridge consisting of oceanic crust/plate), the migration also transports continents to the plate boundaries, which can lead to collisions with other continents. The result is the formation of huge mountains and the growth of continents.

Slices of the thin, hotter crust presumably moved during the entire phase of early crust formation driven by strong convection in the mantle. The processes that we observe today in connection with plate tectonics were certainly not so clearly pronounced in this form back then. This includes the formation of deep-sea trenches along the subduction zones, volcanic belts along island belts, continents colliding with other continents, and much more.

Volcanoes that emerge from the young crust would be the first port of call for models that require a land surface featuring flat pools and the special conditions that prevail here long before continents' cores could develop. However, single volcanoes only emerged at low altitudes. The thin crust with its underlying high-temperature mantle could not support the piled-up masses for long and they literally sank. The amount of ocean water and the depth of the oceans at that time are unknown. The extent to which such volcanoes protruded from the sea cannot be estimated. Only prolonged volcanic activity with an increased supply of changing magmas could have formed more stable crust thicknesses locally. Iceland provides an approximate example in this case. It serves as an example for the development of land masses that succeeded in protruding from the water, while the entire surface of the earth was covered by ocean. In Iceland, the seam runs from the north to the south between the North American and Eurasian plates. Iceland can thank its existence to two overlapping mantle processes. Firstly, it lies exactly on a mid-ocean ridge, on which the constant production of magmas is usually just enough to compensate for the loss of oceanic crust caused by the lateral migration of the plates. Secondly, the production of the ridge is overlaid by activity from a "hot spot," which makes an additional contribution to magmas from a very old reservoir in the deeper zones. "Hot spots" take the form of tubular, mantle regions which have existed for hundreds of millions of years and which feed volcanoes. The origin of the ascending hot mantle material is suspected to be the boundary between the core and the mantle. The most famous example of a "hot spot" is the chain formed by the Hawaiian Islands. The active zone, which supplies Mauna Kea and Mauna Loa volcanoes with magma, lies below the southeastern island of Hawaii (Big Island). At the same time, the plate is migrating to the northwest and taking the volcanic island with it. At some point, the connection to the supply channel above the "hot spot" will be cut off, and a new volcano will begin to develop further to the southeast. The further you follow the trail to the northwest, the older and more eroded the volcanoes are that migrated with the plate. They have long since become extinct and, like the oldest, are gradually disappearing below the surface of the water as the plate as a whole continues to sink with increasing distance from the oceanic ridge.

Since Iceland lies directly on the seam separating two plates, however, it cannot move away as a unit with just one plate. This only takes place on the half which was formed by the respective plate. The central part of Iceland always stays above the "hot spot." This shows very nicely that special conditions could lead to the formation of land masses above sea level even on infant earth. This is an important aspect when discussing some models for the emergence of life that necessitate a land surface. But Iceland does not just act as an explanation for the emergence of land above water.

Its special location also led to the thickness of the crust quadrupling and the formation of modified magmas. Magmas containing granitic or rhyolitic chemism have developed in some exceptional volcanic complexes. Rhyolites are the rapidly cooled volcanic rocks in a magma from which granites are formed in the crust through slow crystallization. They are actually typical of the continental crust, which today is between 30 and 40 km thick (and up to over 70 km thick where mountain roots exist like in the Himalayas). The processes that led to the formation of typical continental crustal rocks in Iceland in the purely basaltic environment of the oceanic crust took perhaps 20–30 million years. They make it clear that the first continental crust may have formed from a purely basaltic environment very early in the course of the earth's progressive cooling. This is important for our examination of the last model. The continental crust is much thicker and lower in temperature than the basaltic oceanic crust. Furthermore, it is chemically or mineralogically much more heterogeneous and provides a wider range of starting materials. Recent work on the development of the continental crust assumes that 25% of today's continental masses were already present 4 billion years ago [15–17].

References

1. Connelly JN, Bizzarro M, Krot AN, Nordlund A, Wielandt D, Ivanova MA (2012) The absolute chronology and thermal processing of solids in the solar protoplanetary disk. Science 338:651–655
2. Canup RM (2012) Forming a Moon with an Earth-like composition via a giant impact. Science 338(6110):1052–1055
3. Ćuk M, Hamilton D, Lock SJ, Stewart ST (2016) Tidal evolution of the Moon from a high-obliquity, high-angular-momentum Earth. Nature 539:402–406

4. Lock SJ, Stewart ST, Petaev MI, Leinhardt ZM, Mace MT, Jacobsen SB, Ćuk M (2018) The origin of the Moon within a terrestrial synestia. J Geophys Res 123(4):910–951

5. Morbidelli A, Lunine JI, O'Brien DP, Raymond SN, Walsh KJ (2012) Building terrestrial planets. Annu Rev Earth Planet Sci 40:251–275

6. Labrosse S, Hernlund JW, Coltice N (2007) A crystallizing dense magma ocean at the base of the Earth's mantle. Nature 450:866–869

7. Tarduno JA, Cottrell RD, Davis WJ, Nimmo F, Bono RK (2015) A Hadean to Paleoarchean geodynamo recorded by single zircon crystals. Science 349(6247):521–524

8. Tarduno JA, Cottrell RD, Watkeys MK, Hofmann A, Doubrovine PV, Mamajek EE, Liu D, Sibeck DG, Neukirch LP, Usui Y (2010) Geodynamo, solar wind, and magnetopause 3.4 to 3.45 billion years ago. Science 327:1238–1240

9. Jakosky BM, Lin RP, Grebowsky JM et al (2015) The Mars atmosphere and volatile evolution (MAVEN) mission. Space Sci Rev 195(1–4):3–48

10. Lammer H, Zerkle AL, Gebauer S (2018) Origin and evolution of the atmospheres of early Venus, Earth and Mars. Astron Astrophys Rev 26:2. https://doi.org/10.1007/s00159-018-0108-y

11. O'Brien DP, Walsh KJ, Morbidelli A, Raymond SN, Mandell AM (2014) Water delivery and giant impacts in the "Grand Tack" scenario. Icarus 239:74–84

12. NASA/JPL-Caltech/R Hurt (SSC-Caltech) (2004). http://www.spitzer.caltech.edu/images/2638-ssc2004-05d-Orbit-Comparisons

13. Hards VL (2005) Volcanic contributions to the global carbon cycle. Br Geol Surv Occ Pub 10:1–20

14. Meschede M, Warr LN (2019) The geology of Germany. Regional geology reviews. Springer, Berlin

15. Rosas JC, Korenaga J (2018) Rapid crustal growth and efficient crustal recycling in the early Earth: implications for Hadean and Archean geodynamics. Earth Planet Sci Lett 494:42–49

16. Rozel AB, Golabek GJ, Jain C, Tackley PJ, Gerya T (2017) Continental crust formation on early Earth controlled by intrusive magmatism. Nature 545:332–335

17. Dhuime B, Hawkesworth CJ, Cawood PA, Storey CD (2012) A change in the geodynamics of continental growth 3 billion years ago. Science 335:1334–1336

3

The Narrower General Conditions: Chemistry, Physics, and Physical Chemistry—We Can't Live Without Them

Abstract The elements which are the main representatives for the development of life are carbon, hydrogen, nitrogen, oxygen, phosphorus, and the subordinate sulfur. All of these elements were available in large quantities right from the start. Phosphorus was only made available by meteorites and minerals from the earth's crust in special circumstances. Reactions that take place between the molecules are subject to various laws of chemistry. Dynamic equilibria, which are accelerated by catalysts, are crucial for subsequent reactions. Besides dilution effects, the increase in entropy and the chirality of the molecules are important factors in organic chemistry.

3.1 The Chemical Resources of Life

One key problem in the discussion about the origin of life is the lack of documents and knowledge of the framework conditions in these early times. According to the planetary conditions considered in the first part of this book, the structure and composition of the cell today and the processes which take place in it need to be used to draw conclusions about actual biochemical development. From this, minimum requirements can be derived that limit the chemical and physiochemical requirements for the origin of life. This therefore makes it necessary to consider all relevant resources and influencing factors in the creation process, insofar as they can be transferred from today to the time 4 billion years ago.

The difficulties that arise in discussing how life is created are directly related to the place where it happens. First of all, organic molecules to start the process need to have been available in sufficient quantities. The elements required to create the simplest form of a living cell are initially limited to carbon (C), hydrogen (H), nitrogen (N), oxygen (O), phosphorus (P), and subordinately sulfur (S). These elements need have been constantly available in the formation environment for long enough for the first larger molecules to have been formed from them. The elements in question are present in different concentrations in interstellar dust/meteorites/comets, the atmosphere, the oceans, and the earth's crust. However, each larger molecule needs its own conditions to be able to form from the initial products. For instance, it is inconceivable that all the components required were available in a narrowly defined environment and that amino acids, organic bases, or sugar could form from them. Even the amino acids in our body have different conditions for formation. For instance, they depend on the pH of the aqueous solution, the temperature, or the ions involved.

However, let us first take a look at the few elements that ultimately stand for what constitutes life, their special characteristics, and how they were represented in and on an infant earth.

Carbon

At the time when the earth first formed, carbon is an element that mainly occurred in connection with oxygen as a gas or dissolved in the interior of the earth. In addition to the reduced form of carbon monoxide (CO), the oxidized form of carbon dioxide (CO_2) also existed. Volcanic activity and escape from the fracture zones in the earth's crust released the gases into the atmosphere from the earth's mantle, where they had a much more prevalent share than today. Information on this from experiments carried out so far varies widely, since there only indirect methods of detection for the level of the concentration exist (cf. Sect. 2.4).

Hydrogen

Under the conditions present on infant earth, hydrogen was dissolved in the earth's mantle and crust or was present as a gas. As soon as it appears in gaseous form, like nitrogen or oxygen, it is always connected to a molecule by two atoms (H_2). With oxygen, H_2 forms a water molecule, a very stable compound that can only be separated with a great deal of energy. The hydrogen was released into the atmosphere through volcanic processes or constant outgassing at fracture zones in the earth's crust. Because of its low mass, it could not be held there for a long time and drifted off into space. Under the conditions in the upper crust of the earth, hydrogen can react with CO and CO_2 to form long-chain organic molecules. These molecules are important building blocks for the development of more complex molecules. A technical variant of this is used in the Fischer–Tropsch synthesis.

Fischer–Tropsch Synthesis

In 1925, chemists Franz Fischer and Hans Tropsch developed a process for liquefying coal at the Kaiser Wilhelm Institute for Coal Research in Mülheim an der Ruhr with the aim of producing gasoline and other long-chain organic compounds. Named after them, the Fischer–Tropsch synthesis formed the basis for fuel production in the Second World War, which needed to be met due to the lack of oil and blockades on supplies of lignite and hard coal. The process has several stages. First, a synthesis gas consisting of carbon monoxide and hydrogen needs to be produced from the coal. The mixture is then subject to pressure of up to 25 bar and heated at temperatures between 160 and 300 °C. The use of suitable cobalt or iron catalysts makes the components react to form long-chain hydrocarbons. Varying the temperature and pressure allows various products to be obtained ranging from fuels to synthetic oils and high-quality organic compounds. In addition to other initial products, such as natural gas or biomass, carbon dioxide can also be used under adapted synthesis conditions.

Haber–Bosch Process

Another large-scale process relevant to the processes in the earth's crust is the Haber–Bosch process for the production of ammonia (NH_3). Fritz Haber and Carl Bosch received the Nobel Prize in Chemistry in 1918 and 1931, respectively, for the results of their research on ammonia synthesis. Ammonia is used as a raw material for various nitrogen compounds that are mainly used in fertilizers. As with the Fischer–Tropsch synthesis, gases are subjected to high pressure and temperatures using suitable catalysts so that the desired reaction occurs. For the NH_3 synthesis, nitrogen is extracted directly from the air and made to react with hydrogen. The process takes place at pressures ranging from 150 to 350 bar and temperatures from 400 to 500 °C using ferrous catalysts.

Nitrogen

Nitrogen is also dissolved in the earth's mantle and was gradually released as a gas when the earth cooled. However, it hardly plays any role in the formation of minerals in the earth's mantle and crust. Under the pressure and temperature conditions in the earth's mantle and crust, nitrogen can react with hydrogen. Ammonia (NH3) is formed, which may have existed in a greater proportion in the structure of the atmosphere than pure nitrogen at the beginning. Similar conditions are also created in the Haber–Bosch process for the large-scale production of ammonia (see box). Hydrogen and nitrogen can also form hydrogen cyanide or hydrocyanic acid (HCN), an important starting material for the formation of organic bases. Under normal conditions, nitrogen exists in the form of gas on earth in which two atoms always form a molecule (N_2). In this form, nitrogen is very inert. Today, it forms 78% of the main part of the atmosphere. It can exist alongside oxygen in the atmosphere without reacting with it. The living environment has a high need for nitrogen. Although the resources in the atmosphere are almost inexhaustible, due to the inertness of the N_2 molecule, nitrogen can hardly be used for biological processes. Certain steps in development were required in the course of evolution which, with increasing need, created the conditions for usability for the living environment.

Oxygen

After iron, oxygen is the second most common element on earth. It forms a major component in most rock-forming minerals, which are built out of a framework of silica tetrahedra (SiO_4 tetraeder). Silicon dioxide compounds are chemically very stable and hardly available as a source of oxygen. Easier to use for the biochemical processes is the water molecule, from which oxygen or the hydroxyl radical (OH molecule) can be extracted for use by various reactions, along with the two compounds of carbon with oxygen, carbon dioxide (CO_2) and carbon monoxide (CO).

Phosphorus/Phosphate

Phosphorus does not exist in pure form, neither inside the earth nor on the earth's surface. It is contained in the mineral apatite in the earth's crust in large quantities. These minerals are so strongly represented in some rocks that they are considered to be rock forming. If they come into contact with hot, acidic solutions in cracks in the earth's crust, they dissolve completely and provide enough phosphate for various reaction steps. Phosphate is a phosphorus-oxygen compound that very easily forms stable compounds with other elements and forms very resistant minerals. Calcium reacts with free phosphate in this way to form the mineral apatite, our mineral from the earth's crust, which is poorly soluble on the earth's surface. But apatite is also the material from which our teeth are made, the hardest parts of our body. In biochemical processes, phosphate is the most important inorganic compound, a nutrient that caters for the algae blooming very strongly when it enters the water with detergents containing phosphate. Phosphate forms the backbone of DNA and RNA and is one of the most important energy sources in the cell (adenosine triphosphate [ATP] and guanosine triphosphate [GTP]).

With one exception, virtually no source of phosphate was originally available on the surface of the earth for beginning life: a characteristic phosphor mineral occurs in iron meteorites, which was discovered 170 years ago by the Austrian chemist Adolf Patera. He named it after the name of the natural scientist Karl Franz Anton von Schreibers in order to honor him. Schreibersite is water-soluble, so that chemical weathering of the meteorites releases phosphorus. Phosphorus quickly oxidizes in the oxygen atmosphere today and in contact with calcium is fixed in a mineral lattice. On infant earth, conditions without oxygen were more complex. Phosphorus reacting with carbon dioxide (CO_2) was able to provide the oxygen required to form phosphate (PO_4^3). This aspect plays a larger role in the discussion concerning the importance of extra-terrestrial building blocks for life. Meteorites may have contributed to the supply of phosphate [1].

Sulfur

Sulfur (S) occurs both in pure form on the surface of the earth and in connection with metals or in the crystal lattice of minerals. It supports many magmatic processes. Volcanic gases bring sulfur to the surface in combination with hydrogen (H_2S) or oxygen (SO_2), where in sufficient concentration, it can precipitate as elemental sulfur. An easily accessible example of this is the rim of the volcano crater in southern Italy, the neighboring island to Lipari (Fig. 3.1). Here the hot sulfur fumes have covered parts of the crater in rich yellow. Large explosive volcanic eruptions carry the sulfur gas high into the atmosphere, where, as an aerosol, it can influence the radiation balance of the sun-earth system in such a way that should not be underestimated. The eruption of Mt. Pinatubo in the Philippines in 1991, for example, caused the Northern Hemisphere to cool by 0.5 °C in the following year. Sulfur plays a not insignificant role in biochemistry. Together with iron, it forms iron-sulfur clusters, which are involved in enzyme reactions. There are also two amino acids that contain sulfur (cysteine and methionine). Cysteine plays a role in the folding of proteins (amino acid chains with more than 100 amino acids)

Fig. 3.1 Sulfur precipitation on the crater rim on the island of Vulcano, Aeolian Islands, southern Italy

by forming special bridge connections in the amino acid chains (disulfide bridges).

In the beginning, each of the elements presented here was available on earth represented in sufficient number, although, as in the case of phosphorus, only under special conditions. Their simple availability does not provide a reason alone for them coming together in parts, reacting with one another and forming complex molecules.

In a figurative sense, the elements are a little like letters which we can use to form words. We do not know how long the words are and what the order of the letters should be or what the words mean, how they should be built into sentences, and what grammar gives the sentences sense. No manual exists that indicates that chapters, pages, and books can be written from the words so that information can be stored or instructions can be provided. So, if we only have the bare letters in front of us in a large collection, basic rules need to be defined that allows us to combine the letters meaningfully. In chemistry, all elements are subject to natural physicochemical laws. They determine the possibilities for forming or breaking connections with others. Understanding these laws is a prerequisite for being able to understand the steps on the path to life, "the words, sentences, pages, and entire books," in the transition from purely inorganic chemistry to organic and biochemical chemistry.

3.2   Chemistry Has Its Own Laws

Chemical reactions between molecules take place in a complex manner, which can be represented in an energy diagram typical for each reaction. In addition to reaction energies, the reaction rate also needs to be considered as another important parameter. In a closed system, a balance always exists between the educts, the starting molecules, and the products, the newly created substances. If a connection between two molecules occurs (whereby the actual number of molecules is always astronomically high for the smallest amounts that we can use), the previously formed connection also disintegrates. Now it all depends on which side the balance is. This is comparable to a beam balance that is in equilibrium, although the weights are distributed very differently on both sides. The balance is achieved by where the fulcrum is located, which is not positioned centrally below the bar. The side with the higher weight has the shorter bar, the other side a correspondingly longer one, in order to create a balance. In the same way, reaction equilibria exist that result in the majority of the reaction product and leave few educts on the other side. Nevertheless, part of the reaction product is converted back into the educts. It keeps going back and forth, with the result that at some point a balance is set from the outside, with a preference for one side or the other. This is therefore not a static but a dynamic balance. The speed of the reaction is of particular importance in the formation of large organic molecules, such as RNA or DNA. Many series of experiments in laboratories have shown that a problem exists in the development of a long strand of RNA under prebiotic conditions that has not yet been solved. An RNA breaks down into smaller strand sections faster than building up longer chains. Conditions or catalysts must therefore be found that enable the development of long RNA molecules, as is the case with today's enzymes in the cell.

3.3   Catalysts Accelerate the Reaction Enormously

In some chemical reactions, the rates of product formation are so slow and the decomposition of the products formed so high that the equilibrium lies almost entirely on the side of the starting products. In other words, they hardly react with each other. Let us take a large bowl whose outer edge has two spheres positioned exactly opposite each other. Each sphere—each represents one molecule—has a very small magnet affixed at a point, one with a positive and the other with a negative pole. We let go of the balls at the same time; they roll into the middle of the bowl quickly and either miss or hit one another. In the latter case, they hit each other hard and repel one another. As a result, they roll part of the way up the edge again, roll back down, and either hit each other again or miss. By slightly swaying the bowl, the spheres remain in permanent motion. After many attempts, the small magnets on the two spheres will eventually meet, and they will stick together. They have managed to connect. We can imagine most chemical reactions in the same way.

In many cases, accuracy can be increased by using a catalytic converter. A catalytic converter is a chemical tool that speeds up the reaction without itself being consumed. It does not shift the balance of the reaction, however. The formation of the products (and their decay) is merely accelerated. The yield can be increased by separating the products from the reaction process. In the case of an enzyme, the catalyst can be thought of like a pipe wrench that holds one of the spheres so that the small magnet is directed outward optimally. This makes the probability that the second sphere with its own magnet will hit the other sphere held by the pipe wrench much higher as a result.

Enzymes have developed into perfect catalysts in nature. These take the form of long, intricately folded amino acid chains that provide pockets for specific molecules. For example, they hold amino acids in an aqueous environment so perfectly that they can be connected with an RNA (e.g., the transport RNA [tRNA]). Also, a connection between two amino acids in water through chance contact hardly ever takes place. The reason for this is that the connection only comes about when a building block is released from each of the two amino acids involved: a hydrogen atom on one side and an OH molecule on the other. In the same step, the two building blocks are connected to form a water molecule and released. The connection then takes place at the positions that become vacant in the amino acids. And that is the problem if the reaction is to take place in water: water molecules that surround the reactants and keep them apart already exist everywhere. They ensure that a release of the hydrogen atom and the OH molecule can only rarely take place. However, this is a prerequisite for the release of the connection positions on the two amino acids so that they can be linked.

The question arises as to how amino acids could react chemically to form longer chains in the early phase of the earth's development, namely, in water, which is assumed to be the main medium for organic chemistry. One possibility would be to exclude the water—at least temporarily, as is the case with the cyclical drying up of shallow pools—or by removing water, as can occur at a mineral interface. The easiest reactions would have been those in an organic solvent. Here the connections take place without opponents, which hinder the release of the water molecule. But where would an organic solvent like alcohol or turpentine come from in the early days of the earth's development? The conditions for the formation of these compounds were initially created by the decomposition of organisms. The path to this was complex. It consisted of the accumulation of biomass in sandy and clayey sediments and their slow conversion under the pressure exerted by subsequent sediment layers and temperatures as deep as several thousand meters, all over very long geological periods. It was only through this, by slowly converting the biomass and then collecting the flowable components in small pore spaces, that petroleum emerged, from which we obtain a large part of our organic solvents today.

3.4 Dilution: No Reaction Without Concentration

The image provided by oil production already clearly shows that biochemical or organic chemical reactions require a high concentration of the components involved to obtain new reaction products during realistic period of time. Suppose a process existed that provided all the building blocks required for the development of life which gradually released them into the ocean. It is easy to imagine that the concentrations of amino acids, bases, and sugars would have had to be infinitely high for the individual molecules to ever meet again in the vastness of the oceans. A process that requires a constant selection of molecules and a constant combination with high concentrations is inconceivable in a free ocean. When the corresponding building blocks of life entered the sea, there were infinitely diluted. No question of the early ocean being a primordial soup can exist. An enrichment of the shallow waters at the edge of volcanic islands or the first continents may represent an alternative. However, constant flooding represented a problem in these zones, because of strong tidal waves and especially after meteorite strikes. As a result, a large proportion of the molecules would have been washed out into the open sea every time resulting in the mass of reactants required being lost.

This means that every plausible model requires the transport of molecules to the place where the reactions can take place. A high concentration with constant replenishment is just as important as the removal of superfluous components. This final aspect has only become clear in recent years. It has been shown that the reactivity gradually comes to a standstill with the constant supply of organic chemical compounds in a suitable space in which the molecules can react. The unusable part needs to be disposed of constantly; otherwise the porridge becomes too thick, and formation processes suffocate.

Ultimately, tar emerges, which originally gave this phenomenon its name [2]. Account needs to be taken of the tar problem right from the start in all considerations aimed at forming the first cell. This means that only those environments come into question that guarantee an open system with the supply and removal of the reaction substances.

3.5   Entropy and Still No End

And then we have "entropy," which is a favorite buzzword of physicochemists and a very important thermodynamic variable. Right at the beginning of the discussion about the origin of life, questions were asked about how entropy behaves in relation to life. Basically, life works against entropy. What did our physicochemist colleagues mean by this, and what is entropy and why is it so important?

Actually, entropy is incredible, and a term that hardly anyone knows. At the same time, entropy is at least as important as energy for all the processes in space and around us. Most of those who have heard of entropy before can't quite put their finger on it precisely, and only a few specialists are really aware of its importance. Entropy is colloquially referred to as a measure of disorder. (Incidentally, parents are very familiar with the process for creating entropy in their children's rooms.) If entropy did not exist, the universe as we know it would not exist either. Right from the start, entropy played a crucial role, whatever things looked like. The universe started with expansion—a process that continues to this day. To put it simply, the disorder in the system that we call space increased and is still increasing. And that's the way it is always, on a large or small scale, in the simplest chemical reactions, in complex physical processes, or in processes that we design ourselves. In total, an increase in entropy has to take place for all reactions. A good example is when water freezes. When water freezes, ice crystals form. Crystallization means that each water molecule takes up a fixed place at a certain distance from its neighboring molecule and is no longer mobile as in a liquid state. A crystal lattice with a high degree of order is formed. This picture is comparable to a crowd of people in a pedestrian precinct prior to a sale or the same number of people who always stand at the same distance from one another on a parade ground in a military parade. Crystallization or order clearly violates the principle of increasing entropy. In order to fulfill the principle of entropy nevertheless, i.e., the necessary increase in entropy during a process, heat is generated and released to the outside when the water molecules take up their positions in the lattice. As a result, the entropy becomes larger in total than it was before the freezing process. Fruit farmers make use of the process of releasing heat when the blossoms on their fruit trees threaten to freeze in spring when a cold snap threatens. They spray the blossoms with water, which crystallizes and releases so much heat that they are not damaged. In order to thaw the ice again, heat energy needs be added correspondingly. As a result of this, the ordered positions in the lattice are given up, and the water molecules are converted into disordered movement as liquid water. Entropy increases here as well.

We also frequently experience this principle ourselves in relation to our own bodies. Tidying up in our home increases the entropy overall, although order is created, and entropy decreases locally in the room. We throw away a lot of stuff which gets disposed of far away from us. We sweat and give off heat. And all this increases entropy. Or take our computer, for example, a large number of processes that create order take place on it when we write an email. Does this increase entropy? We can hear it from the noise the ventilator makes when working: from the heat that is continuously dissipated. Big data centers and server stations have exactly the same problem, the dissipation of heat that swiftly follows the ordering processes that take place in their servers.

In a chemical reaction, there is another way of solving the problem of increasing entropy. Every time two molecules react, order is created. If a large molecule reacts with a smaller one, it can combine with the smaller one by splitting off part of itself. In this case, the split has led to an entropy gain which is slightly larger than the entropy loss connected to the reaction of the two molecules. At the same time, additional heat can be given off.

The examples show that a process takes place in organic chemistry, with the construction of more complex moleculesS, which violates the entropy principle. It can only be avoided by emitting heat (Fig. 3.2) or by breaking

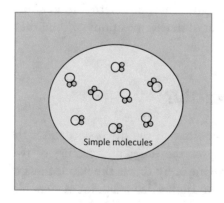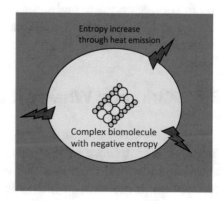

Fig. 3.2 Biological order is generated in a cell by spontaneous heat being released into the environment

down larger molecules formed elsewhere into smaller ones. And an upstream energy source is required at a minimum for heat output. When life started to develop, chemical energy, thermal energy, or potential energy was predominantly available. Only later did biological cells form that could draw energy from existing organic molecules or from sunlight (consumers and producers). Ultimately, everything in a living being is about entropy. All biochemical reactions result in an increase in entropy in the environment and a decrease in entropy in the cells. We need food because it gives us the energy we need to increase entropy in the form of heat dissipation. This is the only way that the molecules in our body can react with one another. If the ambient temperature is too high, the cells can no longer give off heat. The consequence is that there is no increase in entropy and therefore there are no molecular reactions. The cell system dies. This means that starting the process of life also means taking up the fight against entropy.

From today's perspective, as heat-emitting beings, we can actually see that the laws of entropy have been observed on the path to life—one of the essential preconditions for our existence. The other parameters described, such as the concentration of molecules, catalyst-supported reaction rates, or the supply and the removal of components, were also each of decisive importance in their own right and contributed to the development to the necessary extent. All of these factors in their interaction need to be understood in order to understand the origin of life. And this is where another peculiarity of organic chemistry emerges: it concerns the handedness, the orientation of the molecules in relation to a reference system. At first glance, it appears to be a secondary phenomenon, but when you examine it closely, its meaning quickly becomes clear. For example, two amino acid molecules can have two different structures but with the same composition with each causing different properties in complex organic molecules as a result. They are referred to as chiral. The same building blocks are located at different positions within these molecules.

3.6   Chirality: What Is It Exactly?

Chirality addressed one of the major questions in the discussion about the origin of life. It concerns the peculiarity of the different structure of two chemically identical molecules. What we mean by this is the handedness of certain organic molecules, which is different from a defined point of view, like that of your right and left hand. These molecules are called chiral. Determining handedness serves to compare chemically identical molecules that have variations in their structure. The structures behave like a mirror image of an image. Handedness is determined based on the orientation of certain atoms in relation to the central carbon atom. Owing to chemical evolution, certain structures in the formation of complex molecules were preferred in the earliest phase. In chemistry, various approaches exist from history for defining the structure of handedness. Well known, for example, are left- or right-turning lactic acids. Corresponding information for amino acids and sugar also exists. The background for these statements is that when research in this field started, the handedness of a molecule was determined based on the rotation of polarized light. For example, it points to the left for one molecule and to the right for the same mirror image. In the literature on amino acids or sugars, for example, they are labelled with the letter L for left (lat. laevus, left) and D for right (lat. dexter, right). It is worth noting that the D and L variants always occur with the same frequency when the molecules are produced in the laboratory, but one of them (L or D) always dominates when they are formed by biological processes. With a few exceptions, the amino acids in biological cells always adopt the L form and the sugar in the DNA (deoxyribose) or the RNA (ribose) always the D form.

Molecules also exist that only have one structural shape. They are referred to as achiral. The amino acids found in biological cells are all chiral with one exception. Only glycine, the simplest amino acid, is achiral. This exception helps that the probabilities of the combinations in peptides do not become astronomically high. But more on that below when I address the formation of peptides (Sect. 8.3) (Fig. 3.3).

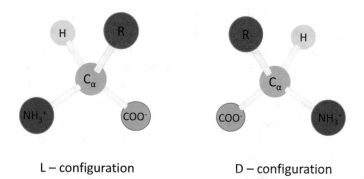

L – configuration   D – configuration

Fig. 3.3 Representation of the L and D configuration in a chiral amino acid. The two molecules cannot be brought into alignment by superimposition. C_α, central carbon atom; H, hydrogen; NH_3, protonated amino group; COO^-, deprotonated carboxy group; R, variable side chain

a   b

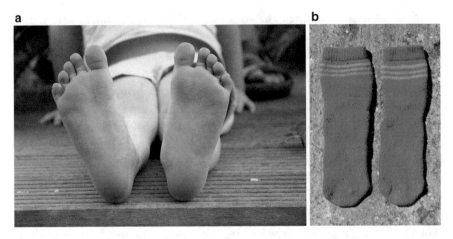

Fig. 3.4 An example of the chiral structure of our (**a**) feet; (**b**) socks are achiral

What exactly does chiral mean? It means that two chemically identical molecules (or products) have different forms that cannot be superimposed. Human hands (hence the handedness) are often used as a comparison. If we look at our palms and superimpose them, the thumb on one hand is on the left and the one on the other hand on the right. The same applies to the direction of rotation of corkscrews for left- and right-handed people or with left- and right-rotating worm castings. We can also take a look at our feet (Fig. 3.4a), whereby we can name a contrasting example, as well our socks. Unworn socks are achiral (Fig. 3.4b).

Why does the preference for a certain molecular structure pose such an important question? If amino acids are formed abiotically, through nonbiological processes, the L and D species (referred to as enantiomers) are always formed in equal parts, regardless of the environment in which they occur. What is referred to as a racemate is formed, a 1:1 mixture of both species. The remarkable thing about this is that the chemical and physical properties of the molecules, whether they are L or D, are completely the same. On the other hand, however, they are biologically very different, since the structure is very important when the enantiomers are incorporated into larger molecules. It is like building a spiral staircase, whose sections should always have the same rotation, in a house with several floors. If the wrong section is delivered, i.e., one that turns in the wrong direction, you can no longer install the staircase as intended in the stairwell. The pain reliever Contergan shows the influence the structure has with the same chemical composition. While in one form Contergan is a safe sedative, in another form it leads to severe malformations of the limbs in offspring when taken during pregnancy.

The second sentence above confronts everyone who enters the discussion about the origin of life with the question about the real significance of chirality. It was the same in all the discussion groups where we met. We used a good dollop of our intellectual capacity on this topic from the start. Which mechanisms have led to only one configuration being established at a time with the same quantity supply and the same chemical behavior of the molecules [3]? The causes and the time when the corresponding form was determined are unknown so far. One of the most recent discoveries is the influence of circularly polarized light in space on the balance of L and D amino acids found in meteorites. The ratio has clearly shifted in favor of the L species [4]. If this is the cause of today's dominant L-amino acids on earth, it would have to be assumed that a substantial proportion of organic molecules from space were available as starting materials for cell formation in the beginning. But a fairly simple solution also exists, which is presented in Chap. 8.

The phenomenon of chirality becomes important when we come to the enrichment and selection of the first molecules for cell development. Both configurations of the molecules always compete with one another. A mechanism which preferred the one direction therefore needs to be identified. Determining a handedness in amino acids is crucial when it comes to the formation of larger molecules such as proteins and enzymes. Although the chemical properties of the two differently oriented amino acids are identical, those with the D configuration cannot achieve much in a world of L-amino acids. An indiscriminate combination of left-handed and right-handed amino acids leads to a disordered chain with different properties than chains with only one handedness of the amino acids. If the chains consist exclusively of the D or L version, folding will occur faster. This in turn leads to more stable structures and thus to a longer lifespan for the molecule. The determination of only one handedness in all cells that originate from LUCA is an indication of the special importance of longer amino acid chains. They have probably played a crucial role from the start. The adoption of a three-dimensional structure represented the precondition for the development of specific catalytic functions, which, as we will show later, have provided crucial support in the formation of complex molecules [5].

References

1. Pasek MA (2017) Schreibersite on the early Earth: scenarios for prebiotic phosphorylation. Geosci Front 8(2):329–335
2. Benner SA (2014) Paradoxes in the origin of life. Orig Life Evol Biosph 44:339

4

Really Helpful: A Brief Outline of What Happens in Biological Cells Today

Abstract To understand the development of life, we need to understand today's processes in the cells that make up the three domains of life. Storage of information probably first took place in the RNA, which was later replaced by the DNA. The code for the sequences of the amino acids in the protein chains is stored in the DNA. It is transferred to an RNA by copying corresponding strand sections, from which the information in the ribosome is used for forming the amino acid chains. The exact assignment is carried out by specifically loaded transport molecules that are linked to the associated amino acid using a specially adapted enzyme.

4.1 The Problem of Containment

From the point of view of science today, the origin of the first self-propagating cell can be viewed from two sides: facing forward, from the formation of the first organic molecules to the appearance of LUCA, and facing backward, based on the knowledge of today's cells to ever more simple forms. In the second case, the functional components need to be reduced to their minimum configuration without losing their characteristic basic function. The second case is comparable to a large city which has grown over the centuries where the original town center is being sought. Even this tiny core had basic structures, such as energy and water supply, and material and information flow. Everything was being constantly developed and kept the growing city complex functional. Each for themselves, the building blocks in today's cells, has been undergoing developments over billions of years, with constant changes, increases, and adjustments taking place. Discovering the functional core of the "molecular metropolises" involves a great deal of effort. But it can be worth it, if it can build a bridge to the other side, the forward facing side. In order to clarify, the most important points about building blocks and reaction steps in our cells are summarized below.

Understanding the processes that take place in our cells for them to maintain themselves so precisely to be able to survive for more than 3.5 billion is not easy. This precision requires so many reaction steps that the foundations that led to the first self-replicating cell are completely obscured. It is like trying to deduce the history of the stone axe from the technology in one of our most modern aircraft. Nevertheless, in all our considerations of the origin of life, we need to understand what is going on in our cells today. They form one end of the line of development. For the other end, the beginning, making simplifications or even using processes from our technical world for comparison is helpful.

If we consider the processes in living bodies, an extremely complicated, finely tuned world of chemical and physicochemical processes opens up, which is still far from being broken down in its entirety. Few experts are able to understand individual steps in the variety of reactions. Nevertheless, basic principles can be recognized using models from the technical world we are familiar with. Everything that occurs in a cell in terms of reaction steps is bound up with inevitably successive reactions, each of which determines the next steps. We can use a toy as a comparison, into which a ball is inserted at the highest point on a track where the specified path takes numerous turns. On its decent, it collides with other balls, which in turn are pushed to the side on new paths and set other balls into motion along different tubes and channels. Switches and mechanisms that trigger special functions are passed ons the way. The whole thing ends when the balls arrive at the lowest level. At this point, external energy must be applied to return all balls to their original position. Any action at one of the contact points or switches can only take place if the sequence of ball movements necessary for further reaction has been carried out beforehand. What happens mechanically in the complicated game box takes place in the cell one after the other through chemical reactions. If a reaction fails—this would correspond to a ball jamming on its way down—the sequence stops and the system dies. This unless elaborate systems of protection are used to maintain them, as they have become increasingly complex in the course of evolution. Based on this raceway model, what we see today is an infinite number of branched tracks along which rolling balls are constantly moving. The energy required to keep the balls at their highest position is covered by the targeted intake of food or, as with plants, solar energy, from the outside.

So far, it has not been possible to draw conclusions about this system from the current number of raceways and their interdependencies, and this seems nigh on impossible. What we are looking for is the first track with the first ball, after which all other tracks developed. At the same time, quantifiable attempts must have been made to combine this first track with ever new variants. Only the individual combination that allowed the principle to be updated or perhaps even improved was retained. But all other approaches were inevitably abandoned.

Getting from this first track to what we can recognize today as the early phase of the three lines of the development of life was still a big step, which is characterized by a deep and seemingly everlasting darkness. Even the oldest identifiable representatives of these lines were so far apart in their development that knowledge of their properties means that little can be inferred about the common ancestors of LUCA. That said, however, we can still compare the genome from different prokaryotes, which is used to form proteins. From this, a path can be seen that points to the basic equipment of LUCA [1]. The data supports the theory that life started in a hydrothermal environment. But more on this in Sect. 8.8.

So, what are these lines of development for life? Cells today can be divided into three domains that belong to two groups (Fig. 4.1). The first group

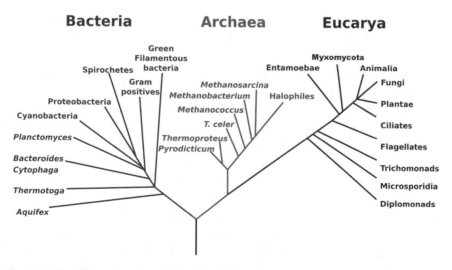

Fig. 4.1 The three domains of life. Figure from hyperlink "https://commons.wikimedia.org/wiki/File:PhylogeneticTree.png," public domain

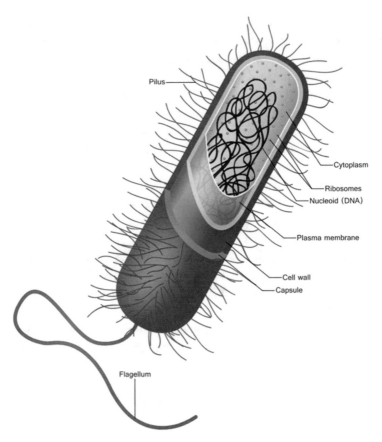

Pilus

Cytoplasm

Ribosomes

Nucleoid (DNA)

Plasma membrane

Cell wall

Capsule

Flagellum

Fig. 4.2 Structure of a prokaryotic unicellular organism. Figure by Ali Zifan, from hyperlink "https://en.wikipedia.org/wiki/File:Prokaryote_cell.svg," public domain

consists of the prokaryotes with the two domains of bacteria and archaea. They are characterized by the fact that their genetic information (DNA) is structured relatively simply and an additional membrane does not limit them separately as a nucleus in the cell (Fig. 4.2). The second group includes the eukaryotes, which simultaneously form the third domain. Its most striking feature compared to prokaryotes is the formation of a cell nucleus in which the DNA is stored (Fig. 4.3) [2]. The eukaryotes originated from the prokaryotes and so play no role in the discussion about the origin of life. LUCA did not have a nucleus yet; it was a prokaryote. Prokaryotic cells have experienced an amazing proliferation. Some of them occur in extreme locations where most eukaryotic cells have no chance of survival. They exist in hot springs at well over 100 °C, at low pH values, in extremely salty waters, and at pressures of over 1100 bar, as evident in the 11,000-m deep trenches found at the

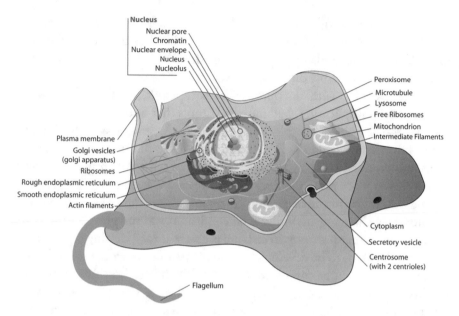

Fig. 4.3 Eukaryotic cell (animal cell). Figure by LadyofHats, from hyperlink "https://commons.wikimedia.org/wiki/File:Animal_cell_structure_en.svg," public domain

deepest points in the oceans. In addition, prokaryotes have also been found in the earth's crust at a 4 km depth. The inaccessibility of these locations is one of the reasons why it is estimated that less than 1% of them are known. Their adaptability allows conclusions to be drawn about how their first ancestors were equipped and the possible environmental conditions in which they existed. This therefore makes them an important indicator for some of the hypotheses concerning the origin of life presented below.

When researching the steps involved in the development of the first cell, the question quickly arose as to what minimum molecular configuration is required for a closed biological shell to be able to multiply itself. The building blocks described below can be defined for this from the knowledge of the process sequences in cells today.

In addition to a membrane that was specially equipped with enzymes, the first cell must have had an information carrier (comparable to RNA or DNA today or with their precursor molecules). Furthermore, a kind of reading device was needed to select the information and convert it into various molecular tools. Today it is the ribosome, a large functional molecule that consists of 32 proteins and 3 RNAs in prokaryotes. It organizes linking the amino acids to peptides using a blueprint of the DNA, which is delivered via a messenger RNA (mRNA). In addition, complex molecular connections were required, which provided assistance during the molecular synthesis and for storing the information. All information on the blocks had to be stored in the information carrier and copied at the same time. This made a doubling of the genetic material possible during cell division.

The cell, which, with the appropriate equipment, took the first decisive step toward multiplication, had molecular tools that were much simpler than they are today. Today's complexity is caused by the exactness with which the stored information from the DNA has to be converted into molecules such as RNA and proteins. It is easy to understand that a cell whose origin is assumed to have taken place more than 3.5 billion years ago is no longer built as simply today as it was in the beginning. The constant pressure of selection owing to constantly changing environmental conditions was too great for this. Even if today's molecular processes in cells are prone to errors, causing sophisticated repair mechanisms to be developed in the course of evolution, the accuracy of the processes cannot be compared to those of the first cell capable of dividing. Perhaps myriad cell divisions were necessary before one of the copies was capable of making a viable version of itself by dividing again. The others simply died. In these experiments that ultimately led to success, the all-important factor was the availability of a resilient information store.

4.2   The Information Store: Without Just Zeros and Ones

The first information store, which emerged from the chemical evolution, is presumed to be in RNA (ribonucleic acid). It consists of a chain structure made of sugar (ribose) and phosphate, with which organic bases are linked in varying order. These are adenine, cytosine, guanine, and uracil. Cells today use DNA (deoxyribonucleic acid) as an information store instead of RNA. It is a twisted, rope-like molecular chain, whose rungs are divided into two parts. The two parts are not particularly firmly connected, and they can be easily separated by higher temperatures (85° C) or by adding alkalis. DNA also consists of four organic bases, of which adenine, cytosine, and guanine are identical to RNA. Uracil takes the place of thymine, which has a slightly different structure than its representative in RNA. The bases of the DNA are arranged in a varying order on both sides of the molecular ladder almost identical to the RNA on a framework made of sugar (here deoxyribose) and phosphate molecules. Only adenine (A) and thymine (T) or guanine (G) and cytosine (C) fit together in one rung of the ladder. If we divide the DNA lengthways in the middle, we get a molecule for each side that is very similar

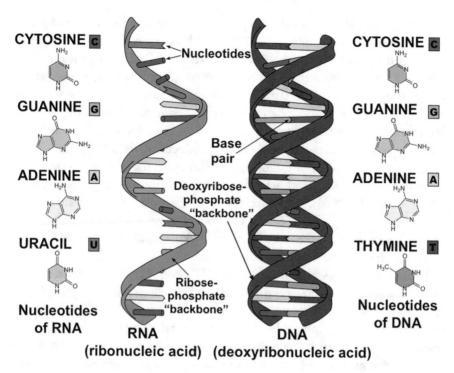

Fig. 4.4 Differences between RNA and DNA. © https://coachingmind.me/uploads/6 0f43595dc46904fd48441a9c90da3e8.png

to RNA. The RNA possesses the two described differences to the DNA: in addition to thymine being exchanged for uracil as already mentioned, a change in the sugar also takes place. The deoxyribose is replaced by the ribose, which is almost identical in structure but has one more oxygen molecule (Fig. 4.4). In comparison to DNA and RNA, the latter is significantly more unstable owing to the slightly modified sugar and for this reason is not particularly suitable for permanent information storage under the conditions prevailing in a cell. Nevertheless, it was able to take over the storage activities required at the beginning—for a cell that only had a short lifespan. There are various indications for the assumption that the development of RNA took place earlier than that of DNA. The molecules from which DNA is assembled are now formed in the cell from the building blocks of RNA. It is therefore likely that even in the early phase of development, the first cells had to be able to build RNA first in order to be able to produce DNA subsequently.

The structural composition of DNA has been known since 1953. X-ray images made by Rosalind Franklin, which were used by the American James Watson and Francis Crick from Britain as the basis for their double helix model, brought about a breakthrough in understanding the storage of genetic information [3]. It also made the differences between RNA and DNA clear at the same time. The presumably later appearance of DNA in an advanced evolutionary step did not mean that the "invention" of RNA got abandoned. In fact, exactly the opposite is true: it continued to be used in a variety of ways, for example, where a short period of survival was not disadvantageous, but sometimes even advantageous. In cells today, different types of RNA with specific codes are constantly required and produced. After use, they are divided up again and recycled. For example, an RNA exists that directly copies information from the DNA and delivers this information to a place of use (messenger RNA [mRNA]), or a specially three-dimensionally structured RNA whose only task is to transport an amino acid from a loading molecule to a processing molecule (transport RNA [tRNA]). More than ten other RNA types have been distinguished in the meantime.

The question arises as to what is so special about DNA as a data repository and, at the same time, how chemical data storage can work. Complex physical data storage is something we are certainly familiar with, at least since the invention of the computer. In the beginning when computers were developed, the punch card was almost like being in the Stone Age. As a storage medium, the punch card allowed light to pass through the holes perforated into it, which a suitable reading device then converted it into a flow of current. Simple texts or sequences of numbers needed entire shoe boxes full of punch cards for smaller operations.

Computer technology made it possible to store columns of numbers, images, or texts using a two-number system, made up of zero and one. In this binary system, all that was required was to define which number and which order of zeros and ones correspond to a number, letter, or symbol. For example, an "A" can be set to 1000001, a "B" to 1000010, and a "C" to 1000011. But the definition alone is clearly not enough. Only a system of switches that is capable of mapping the change between two states can perform arithmetic operations or data storage. The most common form is to apply voltage to the 1 alternating with the 0 state, where no voltage is applied.

In this case, mother nature is a big step ahead of us. She has been using a system built using more than two variables very successfully for more than 3.5 billion years. Information is stored in the DNA by the four bases, adenine, cytosine, guanine, and thymine, or, in short, A, C, G, and T. The four letters offer a greater possibility for variation, and so the data density is significantly higher than with just two letters or numbers. To describe the situation in a more understandable way, we can equate the DNA bases with the numbers 1 to 4. The bases have only two pairing options available to complete the levels on the ladder. A connects only with T and C only connects with G. Applied to the numbers 1 to 4, this means that the sum of the two opposite numbers must always be 5. So only (base) 1 can form a rung with (base) 4 and, accordingly, 2 only with 3. A complex information system can be built from the four numbers, which could have looked like this: the amino acids are linked in the cells in a sequence given by a code to form proteins. The code that is crucial for the position of the respective amino acid in the peptides needs to be unique and guarantee that no other molecule is incorporated into the chain. With the numbers 1 to 4, it could have four digits for an amino acid, so that a maximum of 256 (4^4) amino acids would be used in our cells. That was probably too much of a good thing for the origin of life in terms of complexity, especially since a high number of amino acids like this were certainly not available in the beginning. So no, another way must have existed that was way more complex than the binary system used by computers.

Evolution decided in favor of only three variables from the four numbers. That means that with variation of all possibilities (e.g., 213, 341, 223, etc.), 64 (4^3) unique assignments with three numbers from the pool of the four usable numbers exist at the most (in relation to the bases, three bases always belong to a code as a base triplet (codon) from the four possible ones). Ultimately, with this combination of numbers, 64 different amino acids could be used in the cell system. But this order of magnitude was still too high apparently. The cell system initially worked with less than the 20 recently used amino acids.

In cells today, four codes for start and stop functions are assigned in addition to the 20 combinations for amino acids (canonical amino acids used today) (Fig. 4.5). For example, the start is always begun with the amino acid methionine. At the end of the reading process, however, three different amino acids exist that have a stop function. This leaves 40 of the 64 possible combinations remaining that could have encoded 40 additional amino acids. This was not exploited in the course of evolution either. Instead, double, quadruple, and even six-fold assignments of different codes for individual amino acids exist, since the number combination for 60 cases cannot be fully exploited with 20 amino acids. This means that for the amino acid leucine, six different code combinations (code words) exist alone with which it can be assigned to a chain. A counterpart for each base triplet in the mRNA exists in the form of the anticodon of the tRNA. Only A and U or G and C fit together.

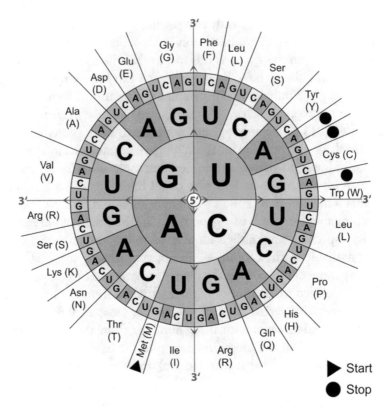

Fig. 4.5 Assignment of the genetic code from the base triplets (codons) for the mRNA to the canonical amino acids (with the bases adenine, uracil, guanine, cytosine, read from the inside to the outside) The amino acids serine (Ser), leucine (Leu), and arginine (Arg) have six different codons, while tryptophan (Trp) and methionine (Met) each have only one codon. The C atoms for the sugar are numbered clockwise from 1' to 5', starting with 1' at the link to the base. In a DNA or RNA strand, the direction of reading for the base sequence can be oriented to this, either from 5' to 3' or in the opposite direction (© Mouagip https://de.wikipedia.org/wiki/Datei:Aminoacids_table.svg)

4.3 How Is the Information Stored Converted in the Cell?

The process of transferring information from the DNA is linked to complex reaction steps by specially adapted molecules. But this does not yet play a role in the basic steps in the development on the pathway to life. Nevertheless, it is helpful to look at the basics of the reactions taking place today, since processes have steadied and molecules are used whose simple precursors formed the basis for later processes.

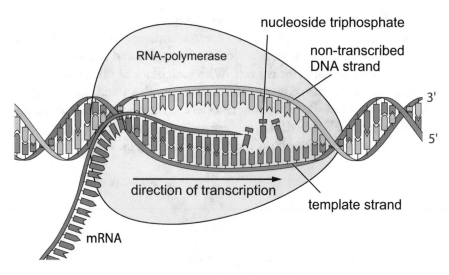

Fig. 4.6 Selection process for partially untwisted DNA. Creation of a messenger RNA (mRNA) that adopts the information by successive attachment and subsequent linking of nucleotides (© Springer-Verlag GmbH, modified [4])

To select the information stored in the DNA, it is untwisted momentarily and divided in length so that the information can be copied from a longitudinal section. The copy is made by attaching complementary RNA building blocks, or nucleotides, to those from a partial strand of the DNA (Fig. 4.6). The division into blocks of three (codons) is carried out in exactly the same way as specified by the DNA. This becomes understandable again when compared with a combination of numbers. A series of alternating numbers between 1 and 4 is supplemented with numbers on the opposite side so that the sum always results in 5. The opposite row corresponds to messenger RNA (messenger RNA [mRNA]), which transports the information to a molecular tool (ribosome) that generates peptide formation. If we look at the bases, a small variation exists, since a base in the DNA (thymine [T]) has a slightly different composition than its counterpart in the RNA (uracil [U]). Position G in the DNA becomes C in the RNA and vice versa. A in the DNA becomes U in the RNA, and T in the DNA (which is replaced by uracil U in the RNA) becomes A in the RNA.

DNA: G C A T
RNA: C G U A

The mRNA migrates to the ribosome, the molecular tool for peptide formation, through which it passes in a certain sequence of steps. This is the most important process in the production of enzymes and proteins, the tools and the building material for the cell. With each block of three (codon) from the bases, the mRNA provides an information unit that stands for exactly one amino acid. And now the following takes place: the ribosome allows another RNA, the transport RNA (tRNA) with an amino acid that is precisely matched to the code, to position itself on the block of three (Fig. 4.7). The transport

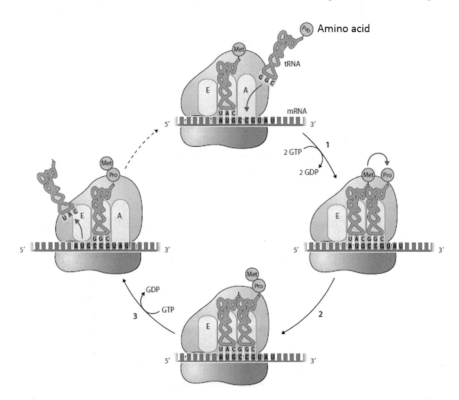

Fig. 4.7 Linking amino acids in the ribosome after the positioning of specifically loaded tRNAs on the codon in the mRNA strand. The first tRNA is placed in the ribosome at the start codon in the mRNA. Every process always starts with the same tRNA amino acid unit, methionine (Met). The next unit (Pro) follows that matches the code for the mRNA (Step 1). The second amino acid is linked to the first (Met) (Step 2), and then the ribosome moves one codon unit to the right (movement from 5′ to 3′, see Fig. 4.5), which creates space for the next loaded tRNA, and the discharged tRNA from Met is released (Step 3). A on the right side corresponds to the position where the loaded tRNA is taken up, and E corresponds to the position where the discharged tRNA is rejected. With each subsequent tRNA on *A*, a new amino acid is bound to the chain, exactly as specified by the information from the mRNA. GTP (guanosine triphosphate) provides the energy required by splitting off a phosphate molecule. This turns it into GDP (guanosine diphosphate) (© Springer-Verlag GmbH [4], Fig. 11–31 modified)

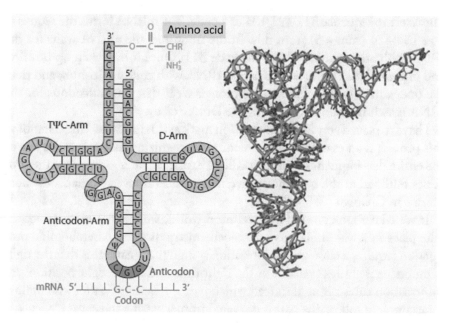

Fig. 4.8 Transport RNA (tRNA) in two and three dimensions. The amino acids are linked on the strand (ACC) protruding upward in the two-dimensional representation. This "trailer coupling" was selected at the beginning when life started to develop and is still the same today for all tRNAs and can be used for all amino acids. It is unspecific, so that the precise assignment of the tRNAs to the associated synthetases ("loading stations") required takes place over other recognition sections on the tRNA. The tRNA is characterized by the alternation of double-stranded sections with loops that contain modified standard bases (TΨC arm, with T for thymidine, Ψ for pseudouridine, and C for cytidine; D arm with the base dihydrouracil). The base triplet is located on the anticodon arm, the information unit as a complementary counterpart (anticodon) to the codon for the mRNA (© Springer-Verlag GmbH [4], modified)

RNA is the most important component in the whole process. At one end, it carries the anticodon, the code, and, at the other, the associated amino acid (Fig. 4.8). The positioning only takes place if the information unit from the three bases, the codon, matches the counterpart, the anticodon from the tRNA, exactly. In numbers, this means that all three sums add up to 5: 214 matches 341, 444 matches 111, etc. Each tRNA has its own code and its own (specific) amino acid. The connection in the ribosome is only temporary. It lasts until the amino acid has separated from its transporter and linked with the amino acid from its predecessor. This creates a chain that is formed by continuously combining mRNA and tRNA.

To make this understandable from another perspective, the situation can be represented by a column of numbers starting with the DNA. The DNA untwists the sequence 311 421 131 as a code. The mRNA forms the sequence 244 134 424 (sum = 5) from this. In the ribosome, the mRNA waits for the tRNA with an amino acid with the code 311. Once it has taken up position and unloaded the amino acid, the next tRNA with code 421 comes, and then the one with code 131 follows. It becomes clear that the anticodons for the tRNA match the number code for the DNA exactly.

I have to place a very large exclamation mark at this juncture. Understanding this process is an essential requirement for recognizing that this is the key to the entire development of life as it will be explained later—only that the processes outlined at the beginning have taken place in reverse order. But more on this in Chap. 8.

If we take a look around our technical world, we see that many processes take place in a way similar to how biochemistry dictates. Assembly line production requires a close-knit information system that guarantees that the right component is always available at the right time and in the right position. An information carrier is used to read which part is required next for assembling, for example, a robot. The part is delivered from a passing transport trolley and placed on the belt. There it is immediately installed into the already built part of the robot. The trolley returns to reload and return back to the line after a short while to be unloaded when requested.

As explained above, in spite of the 60 possibilities given for an exact assignment of codons from 4 different bases in the course of evolution, only 20 amino acids were selected in most cases, which, today, stand for the infinite variety of enzymes in a cell. The information from the DNA transmitted by the mRNA is used accordingly for the formation of chains in which all 20 amino acids normally occur. Their number varies and can be in the thousands in peptide and protein chains. And you can see immediately here why evolution stopped at "only" 20 amino acids in order to try out variations in the chains for forming functional proteins. The number of different combinations of 20 amino acids in a chain of 60 units already results in a size that corresponds to the number of atoms that occur in space. The number of possible variations in a protein with 1000 or even 10,000 units can no longer be quantified. These figures make it clear that there are so many combinations already possible with the existing 20 amino acids that they can never be fully exploited. It also shows that the development of an information system could only commence by beginning with a minimal number of participating molecules and accompanied by an effective selection process.

The sequence of the amino acids and their order in the chains are determined by the DNA code today. In the beginning, it was probably just the RNA. It was subject to changes that were mainly caused by mutations from the beginning.

At this point, the question remains as to how the tRNA is loaded so very precisely, so that exactly the amino acid that the mRNA code specifies is incorporated into the chain. If an error occurs here, the entire configuration of the chain changes immediately. Given the complexity of the folded peptides, it is easy to understand that this can result in the formation of other structures which, for example, impair an enzyme's catalyst properties. If this affects enzymes that control important metabolic processes, cell damage or, in extreme cases, cell death can occur.

The lack of utilization by other amino acids of the 40 code words still available has led to some of the canonical amino acids used today being able to use several transporters for their transport. In addition to single assignment, amino acids that can specifically occupy two, four, or even six different tRNAs also exist. Surprisingly, however, a preference for the use of a very specific tRNA has developed in species today, although five more are available as is possible in some cases. The question is how this assignment is organized when the requirement for reliability is high. Large enzymes, the tRNA synthetases, which take on this task have been developed for this purpose in the course of evolution. These take the form of special molecular tools that ensure that only one amino acid is linked to the corresponding tRNA. Each amino acid species has its own synthetase which only merges it and one of the matching tRNAs.

And so, the circle closes. The synthetases load each tRNA with a special amino acid very precisely. The transporters bring the amino acids to a kind of zip system, where they are linked to form chains based on the information from the DNA. Depending on the function they have developed, these peptides or proteins fold and form a large part of the molecular tools necessary for a cell to function. This also includes enzymes that catalyze the formation of DNA, RNA, ribosomes, etc.

Amino Acid Chains

Amino acids can be linked together to form chains. They are all made up of the same basic molecule, which, depending on the amino acid, is expanded by the associated attachments, what are known as side chains. The same basic structure allows the molecules to always be connected at the same positions with splitting off of a water molecule. For this purpose, an OH group is added from one side of the first amino acid, and an H^+ ion contributed from the second amino acid. The connecting points which become free combine, and a dipeptide is created from two molecules. With further growth, a tripeptide, tetrapeptide, etc. or generally an oligopeptide of up to ten units results. Polypeptides are connected in a chain with higher quantities. Another limit is reached at around 100. The longer chains below are called proteins. The chains can twist and form an alpha helix. It is a common secondary structure that gives the protein the greatest stability. In addition, beta sheets appear as a further secondary structure, or more complex structures are formed from the secondary structures, which become tertiary structures. The folded proteins in turn can assemble and form giant molecules (quaternary structures with up to 30,000 units in the largest proteins). This is how most enzymes are created which have important functions in the cell processes. However, some enzymes also exist that are not made up of proteins. These include the catalytically active RNA (ribozyme), which supports the amino acids linking in the ribosome.

The exponentially increasing number of combinations with the participating 20 amino acids is important. The combinations are calculated based on the power of 20^n. This means that with the possible use of 20 amino acids, a chain of four can already adopt 160,000 different combinations (20^4). If we delve into the world of protein, we start at 20^{100} variations. This is such a vast number of possibilities that nature would not even be able to try out all of them in billions of years. Nevertheless, some combinations have prevailed and are stored in the DNA. This makes it clear that catalytic properties can be developed very easily from a certain chain length. If these properties were rare, the large number of combinations would make it difficult to find suitable functional molecules, even during the long periods of time. It becomes apparent at the same time that at the beginning, only a very few amino acids could have been involved in the interaction between peptides and RNA storage, so that the possible combinations did not become too astronomical. For example, if only two amino acids are combined with one another, already a billion different possibilities exist for a peptide with 30 units (2^{30}). It is not unreasonable to assume that functional molecules could already be formed as a part of this.

And this is exactly where the real problem arises when discussing the origin of life. We refer to it figuratively as the chicken and egg problem, which poses the question which came first, the chicken or the egg. The problem is how the information about the structure of the enzymes and especially the synthetases got into the DNA or at least its precursor, the RNA. Synthetases are molecules that can only act as catalysts from a certain size. The exact order of the amino acids required for this cannot have arisen by chance, like the example calculation with the large numbers shows. Especially not with 20 different species. However, synthetases are absolutely necessary to ensure an exact assignment of the respective amino acids to the code from the information store. This means that they should already need to be available to be formed for the first time. I will try to answer this most important of all questions in Chap. 8.

References

1. Weiss MC, Sousa FL et al (2016) The physiology and habitat of the last universal common ancestor. Nature. Microbiology 1:16116. https://doi.org/10.1038/NMICROBIOL.2016.116
2. Hug LA, Baker BJ, Anantharaman K et al (2016) A new view of the tree of life. Nature. Microbiology 1:16048. doi.org/10.1038/nmicrobiol.2016.48
3. Watson JD (1968) Double helix. A personal account of the discovery of the structure of DNA. Athenaeum, New York
4. Fritsche O (2015) Biologie für Einsteiger. Springer, Berlin

5

The Previous Models: Sighting the Great Nebula

Abstract Over the last hundred years, the question of how life came into being has developed into ideas that encompass a wide variety of environments. At the same time, the first scientific experiments began to detect the formation of organic molecules. The best known is the experiment by Urey and Miller, who obtained corresponding molecules from suspected components of the early atmosphere with the help of lightning discharges. In addition to an iron-sulfur world, which was heavily discussed after the discovery of black smokers and white smokers, the model for pools that cyclically dry up has received increasing attention in recent years. But the formation of organic molecules and life in space has also played a role in the discussion from the start.

5.1 From Ancient Times to Modern Science

The information provided in the chapters above is the minimum required for a further discussion on the origin of life. This makes it clear that only a group of scientists is capable of covering all the areas of the scientific disciplines involved in depth to some degree based on latest findings. The most important thing here is that the members of the group are only able to assess the basics of neighboring disciplines to the extent that they contribute impulses and related considerations concerning the overall question. Understandably, these favorable conditions as I have described them for the Essen Group could not have existed in earlier times. The scientific foundations alone were

© Springer Nature Switzerland AG 2020
U. C. Schreiber, C. Mayer, *The First Cell*, https://doi.org/10.1007/978-3-030-45381-7_5

missing. As in science as a whole, a development took place concerning the question of the origin of life, which led from nonscientific ideas to careful attempts at explanations by individuals to laboratory experiments by smaller research groups.

In ancient times and in the Middle Ages, the scholars had the idea that life could arise spontaneously from earth or mud. The person who established this idea was the Greek philosopher and naturalist Aristotle, who propagated spontaneous generation of life from inanimate matter. But this was a fallacy given rise to by observations that were incapable of looking into the matter deep enough owing to the lack of technical aids. The development of worms and larvae in muddy environments or mold on food was known to everyone at the time. The knowledge available at that time prevented access to proof of the propagation pathways because bacteria or spores could not be perceived because of their minute size. The situation in this respect did not change until modern times. Thus, for centuries there was no reason to doubt the interpretation of one of the best known scholars in antiquity. The idea even still existed in the nineteenth century that, in addition to the known ways of reproduction (biogenesis), life could be propagated at any time, through a variant known as primordial production (abiogenesis). During this time, however, the discussion on the primordial generation of life was no longer so much about the fundamental question of the origin of life but about the additional possibility of the spontaneous origination of the new in parallel to existing life. The research carried out by the French chemist, Luis Pasteur, in the second half of the nineteenth century put an end to this discussion. He was able to demonstrate the influence of microorganisms on fermentation and mold growth through experiments on fermentation and sterilization processes. The main focus had been on these during the late phase of the dispute on spontaneous generation.

5.2   Modern Beginnings

Scientific consideration of the question of the origin of life started only hesitantly in the last third of the nineteenth century, at the time when the British naturalist Charles Darwin formulated his thoughts on this. He assumed a warm little pond as the location for the place of origin of life, in which the development of more complex molecules began with the presence of sufficient inorganic compounds and energy. Life was meant to have developed from these ultimately. This process is no longer applicable today, however, since biological activity would immediately destroy all attempts to start over.

However, it was not until the twentieth century that Soviet biochemist, Alexander Ivanovich Oparin, born in 1894, laid the actual foundations for research into prebiotic evolution. He formulated a hypothesis in which he made it clear that the initial conditions in infant earth differed significantly from those of today. He made statements about the composition of a primordial atmosphere; postulated lightning discharges, sunlight, and volcanism as a source of energy; and coined the term primordial soup for the collection of all the resulting molecules in a primordial ocean [1]. His considerations are now out of date since a different composition is assumed for the primordial atmosphere. In addition, due to a lack of knowledge, the planetary conditions could not be taken into account, and the molecular concentrations in the ocean were far too low for a reactive primordial soup. However, he did lay the theoretical foundations for one of the most well-known experiments in science, which went down in scientific history as the "primordial soup" experiment. This was an experiment performed by the American chemist, Harold Clayton Urey, and his doctoral student Stanley L. Miller, in the 1950s. He built on the hypothesis from Oparin which led to the formation of amino acids directly from simple inorganic components in the laboratory.

5.3   The Experiment by Harold C. Urey and Stanley L. Miller

So, why didn't we just perform an experiment with the potential to support Oparin's considerations? Urey and Miller must have asked themselves something similar at some point in the early 1950s before they began to devise an apparatus in which some of the supposed conditions on infant earth could easily be simulated (Fig. 5.1) [2]. They transferred various gases, which they assumed were characteristic of the primordial atmosphere, into a flask system that was partly filled with water. These were comprised of ammonia (NH_3), methane (CH_4), hydrogen (H_2), and carbon monoxide (CO). The water was heated in the flask so that water vapor rose, and circulation started. Electrical discharges added energy to this mixture, a process that was thought to correspond to lightning striking. And a great surprise ensued. Amino, carbon, and fatty acids formed relatively quickly from the inorganic substances and methane, which represent an important factor in organic chemistry and biology. The sensation that resulted was perfect. From then on, it was believed that the molecules formed had been washed out of the atmosphere into the oceans, where a "primordial soup" had developed in which everything else on the path

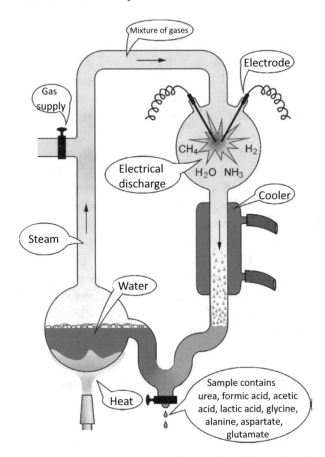

Fig. 5.1 Experimental setup by Urey and Miller for the abiotic synthesis of organic molecules (© Springer-Verlag GmbH [3], modified)

to life took place. It seemed only a matter of time before knowledge about the development of the first cell would come to light.

But criticism was quick to follow. The first thing that became clear is that the concentrations of the molecules in this type of formation were so low that they would have decayed before they could have collided with another molecule in the ocean. In addition, the idea about the composition of the atmosphere also changed. Miller and Urey assumed a reduced atmosphere, visible from the CO and NH_3 in the experiment. However, the outgassing of the earth probably mainly produced CO_2 and N_2. There was also the problem with chirality. In the experiment, the left-handed and right-handed amino acids formed in equal proportions. In nature, only left-handed amino acids occur with a few exceptions. Connecting the results of the experiment to cell formation or even to an RNA was not possible with this experiment. Even though much of the criticism was justified, it ultimately represented the breakthrough everyone was waiting for in research into the origin of life. The experiment showed that a way existed to tackle this seemingly unsolvable question with a series of experimental steps for the first time.

5.4   The Dam Was Broken

A large number of experiments were carried out subsequently to tackle the problems of linking amino acids to peptides or connecting nucleotides to RNA strands. Studies of clay minerals showed that their mainly negatively charged surfaces offer effective contact with positively charged organic molecules and can serve as catalysts in the connection [4]. This also happens on iron disulfide pyrite (FeS_2) which is reduced in contact with hydrogen. This means that a sulfur atom is removed from the connection with the iron and connects with the hydrogen. This process also provides energy that can be used to link the amino acids together, for instance. Pyrite is a mineral that was extremely common on infant earth. The research on this is linked to the name of Günter Wächtershäuser, a patent attorney from Munich who developed an alternative scenario in the 1980s for the early evolution of life, but not without extensive criticism [5, 6]. His hypothesis on biogenesis differed significantly from all models discussed so far. While the hypothesis for the RNA world (cf. Chap. 6), for example, requires a system for the initial phase of life that maintains itself by continuously copying chemically stored information, for Wächtershäuser, the metabolism was the most important factor. He assumed that the formation of organic molecules with the carbon from CO_2 or CO required a mineral surface that was available as an electron supplier. The focus of his discussion revolved around the formation of iron-sulfur minerals, in which the mineral pyrite (FeS_2) occurs. In this way, FeS_2 can be oxidized from an iron-sulfur compound (FeS) in contact with hydrogen sulfide (H_2S). This process releases enough electrons and H^+ ions, which are said to have been directly available to adherent CO_2 or CO molecules for building more complex molecules. His considerations cannot be dismissed entirely out of hand. Some of the enzymes today do have so-called iron-sulfur clusters. There, they adopt important catalytic functions, whereby they transform the substances that were available at the very beginning of development, such as H_2, N_2, or CO. But Wächtershäuser did not get much further with his considerations. The link to enzyme development and the formation of RNA was missing.

5.5   Black Smoker: A Parallel World

Wächtershäuser's hypothesis was interesting in connection with the discovery of black smokers in 1979 off the Galapagos Islands, which so to speak represent an iron-sulfur world (Fig. 5.2). Black smokers are exit points for up to over 400 °C deep sea springs, from which an abundance of dissolved metallic compounds rises from the cracks in the crust, reacts with the seawater, and forms metal sulfide, oxide, and sulfate deposits. This creates regular clouds of black, finely dispersed masses, some of which form to create chimney-like structures after sinking. The hydrothermal vents are part of active ocean ridges where two adjacent plates drift apart. Magma from the underlying earth's mantle rises along the seam between the plate boundaries and supplies the energy that keeps a huge circulation system alive. On the sides of the active ocean ridges, seawater seeps into the depths through the crevices in the thin fractured crust. Here it heats up and reacts with the minerals in the basalt, the

Fig. 5.2 Black smoker—summit region of the active North Su volcano in the eastern Manus basin at an approx. water depth of 1200 m (© MARUM, Center for Marine Environmental Sciences, with kind permission)

rock that exclusively forms the oceanic crust. Metals that react with sulfur are extracted from the minerals in particular and transported away. The circulation closes with the ascent of these hot, mineral-laden waters to the black smoker's exit points.

A highly specialized ecosystem has developed in direct connection with black smokers, whose food base is one of the very rare exceptions that is not sunlight. The entire system is based on the chemical energy provided for chemolithotrophic bacteria and archaea by the escaping metal sulfides. The microorganisms use electrons from redox processes to control chemical reactions for their metabolism. They acquire the carbon necessary for their own cell components from the inorganic carbon from CO_2 and CO. Building on this, a food chain has developed that includes worms, crayfish, and other higher animals. The discovery of these highly specialized unicellular organisms quickly led to the consideration that black smokers could be a model for the origin of life [7]. The thermophilic microorganisms seemed to form an ideal link to those that later, in the course of evolution, probably specialized in cooler environments. They are the simplest organisms we know. This also made it clear that the depths of the ocean provided protection from solar wind and UV radiation. Even large meteorites would have only partially evaporated the water, meaning that the continued existence of the developing world would not have been endangered. However, objections ensued because the chimneys only have a short lifespan of a few years and as a result do not provide long-term conditions for the requisite periods of time. Furthermore, it became clear that the solution load along with the metals and the temperatures in the ascent paths is too high, so that hardly any organic molecules such as nucleotides can be formed. At high temperatures, such molecules disintegrate faster than they are formed. If under certain fringe conditions they still form with the help of mineral surfaces, any further connection to the development into the first biological cell cannot be discerned. Linking amino acids to peptides in water is also very problematic. The connection of two amino acids takes place by splitting off a water molecule, which is formed from an OH molecule from one amino acid and a hydrogen atom from the other. If the amino acids are in the water, they are shielded from one another by water molecules so the reaction cannot take place. The connection only works in the aqueous environment of cells because enzymes provide the appropriate help. In the end, approval for the black smoker model waned more and more, and it was no longer pursued.

5.6   A New Discovery: The White Smoker

The task of the black smoker as a model case for an area in which the development of life originated ran almost in parallel with a discovery that took place on a research trip in the mid-Atlantic at the turn of the millennium. A new hydrothermal region was discovered in the year 2000, which has significantly different properties than all previously known hot springs on the seabed. It lies in a tectonically active zone on the mid-ocean ridge, on the edge of a submarine massif made of mantle rock (Lost City in the Atlantis Massif). This is the special thing about this location: as in most other cases, it does not comprise a large volcanic complex, but a section of the upper mantle raised by constriction [8]. The composition of the rock consists of minerals that are typical of mantle rock. They are very rich in magnesium and iron, but low in silicon. The prevailing minerals olivine and pyroxene are converted into new, water-rich serpentine minerals (serpentinization) by circulating, warmed water. This led to the release of hydrogen and methane. Limestone towers up to 60 m high have formed on the ocean floor, from which water with low temperatures of up to 90 °C emerges—a clear difference to the temperatures of black smokers of over 400 °C (Fig. 5.3). In contrast to the black smokers, white smokers do not have high metal loads. Dissolved calcium compounds emerge with the water from the crevices and channels in the limestone towers that allows them to continue to grow in contact with the CO_2-rich seawater. White smokers have existed for at least 300,000 years and are therefore much more durable than the black smoker springs. Methane, hydrogen, sulfide, and hydrogen accompany the carbonate-rich solutions and provide a livelihood for bacteria and archaea, which in turn form the basis for a complex food chain. However, the wealth of life that forms in white smokers is much less than in black smokers. The lower temperatures in these springs and the perforated limestone towers provide reason to define this milieu as an ideal environment for the emergence of life. In the years that followed, in part until today, some researchers therefore focused on the white smoker model, which they assumed existed at comparable location in the oceans on infant earth [9]. However, fundamental problems still exist here. Besides the likewise problematic environment of the water, even more decisive are the high pH values of the alkaline fluids, which lie between pH 9 and 11. They are a death knell in the argument for the development of RNA. It decomposes rapidly hydrolytically at high pH values.

However, further arguments exist to question the "white smoker" theory as determining the locations where life originated. They are a recent product,

Fig. 5.3 Special smokers in the summit region of the active North Su volcano in the eastern Manus basin off Papua New Guinea at an approx. water depth of 1200 m. These are "smokers" from which liquid sulfur and bubbles of CO_2 escape. With a pH of 1.4, the water is very acidic (© MARUM—Center for Marine Environmental Sciences, with kind permission)

formed under current conditions in the ocean with the current composition of ocean water. It must be assumed that due to high levels of dissolved carbonic acid and sulfur compounds, the ocean water had a significantly lower pH value in the earth's early days. The solubility of lime increases significantly in such an environment, so that the formation of long-lasting lime towers at the hydrothermal exit points is highly questionable. The sunken plateau in the vicinity of the white smoker most probably had a coral reef. The location within the 30-degree zone north of the equator allowed for reef formation due to the water temperatures. The proximity to the water surface of the complex, which was probably higher in earlier times, must have been given, as a profile of the white smoker site by Kelley et al. [10] shows (Fig. 5.4). This shows that the limestone chimneys are underlaid by a talus (limestone debris from a coral reef) and a sedimentary succession of limestone. This situation may have favored the formation of the chimneys and is inconceivable for the early phase of earth. The latest finds at exit points in the eastern Manus basin, which

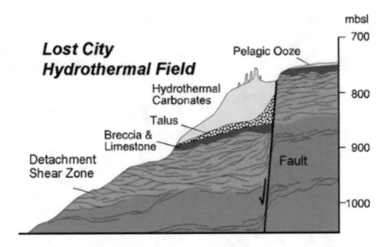

Fig. 5.4 Geological profile of the hydrothermal field lost city with "white smoker." The underlying limestone layer sequence (breccia and limestone) and debris from a coral reef (Talus) are clear. It overlaps a tectonic shear zone in the upper area of the mantle rock. *mbsl* meters below sea level. Figure from Kelley et al. [10], public domain

resemble white smokers (research trip by FS Sonne SO216, 2011 [11, 12]), show extreme conditions with liquid sulfur escaping, which are unsuitable for the development of life.

5.7   The Search Continues: Warm Ponds

Charles Darwin was the first to come up with the idea that life may have started in a warm pond. In a letter to botanist Joseph Hooker in February 1871, he speculated: "But if (and oh what a big if) we could conceive in some warm little pond with all sorts of ammonia and phosphoric salts, light, heat, electricity etcetera present, that a protein compound was chemically formed, ready to undergo still more complex changes" [13].

This idea has been taken up recently since it apparently helps to solve one of the main problems. Wet-dry cycles on land make it possible to link amino acids to peptides that do not function in water without the existence of enzymes. The proponents of this model are supported by the discovery of organic molecules in meteorite rocks, in which a variety of different amino acids have been found, including those that are not found in biological cells. This is an important indication that the finds do not represent contaminations that can take place very quickly when contact is made with the atmosphere and ground.

The scenario is shaped as follows: long after a solid crust had formed, meteorites and cosmic dust rained down on a volcanic island complex raised above sea level, bringing with them organic molecules originating from space. Some of the molecules survived the impact and found their way into the rainwater owing to the weathering of the meteorites. Streams that flowed from the volcanic mountains into the foothills fed lakes and ponds, some of which dried up seasonally and then flooded again. Some of them originated in hydrothermal systems, which added organic compounds. Both muddy and shallow pools in the vicinity of volcanically active regions serve as an example of this (Fig. 5.5). Starting out from these pools, a connection was established to the sea through rivers [14]. Also not excluded is formation of additional molecules on land under the influence of UV which contributed to the diversity of the offering.

The ponds drying up created the moment when the amino acids could react to form peptides. The continuing steps are complex and have not yet been fully formulated. Cells which formed the first bacterial colonies (stromatolites) as they continued to flow into the sea are said to have developed within a very short period of time. The time span for this appears to be extremely short, considering the millions of years required to obtain decisive changes in the development of later cells. In later times, however, a functioning system of processes that ensured the cells' self-preservation already existed.

Fig. 5.5 Example of a mud pool connected to a hydrothermal system (Iceland)

It is very likely that the initial phase up until the first cell required a great deal of time for the infinite number of trials using the trial-and-error principle. No consensus on the time frame exists, however, because the steps on the path to life lie too far back in time. Only the time period for individual steps in the reaction can be specified. Only one thing can be determined from the many attempts required to exploit the range of possibilities available: a great deal of time was needed. And it cannot even be taken into account here that a development that was on the verge of a breakthrough was wiped out by a global event. After this event, everything started over again.

The geological processes that took place on the first volcanoes to appear above sea level were anything but slow. Acid rain strongly affected the unprotected rock since neither soil nor plant coverage protected the rock surface. The result was intense weathering, which led to high erosion rates caused by heavy rainfall. This resulted in a high sediment load in streams and rivers, which filled every pool and small pond in a very short period of time. The pools with a low absorption capacity that dried up were filled after a few hundred years. If arid zones, similar to Australia, existed on the first continents, pools that dried up must have been quickly filled with salts, carbonates precipitated out of solutions, and volcanic ashes from the ubiquitous volcanoes. Corresponding examples are still widespread in arid regions. If new pools were formed, they were not linked to the processes that took place in their predecessors in most cases, so it was like a new start. However, the space required for the development of life needed to have been available for millions of years.

Other arguments exist against the pool model. As the moon is so close to the earth, the ocean tides were extremely strong shortly after the earth formed. The water masses that smashed against the island mountains in the ocean resulted in much faster erosion than is the case today. Not infrequent impact events also took place, which statistically speaking predominantly took place in the sea and presumably washed over the islands completely. The situation was different for larger complexes like Iceland or the first small continents, which were spared the major floods. In this case, the factors mentioned at the beginning need to be discussed in relation a model based on the unprotected surface of the earth as the point of origin. During the initial phase, the sun radiated at around 30% less, but UV radiation was stronger by several orders of magnitude. The reason for this was the initially high rate of rotation by the sun, which produced a strong dynamo effect with increased activity on the sun's surface [15]. Since there was a lack of oxygen in the atmosphere to the development of photosynthesis, no ozone layer existed to block the UV radiation. As a result, much stronger radiation hit the earth's unprotected surface during the period of the first billion years. Possible formations of long-chain connections were therefore exposed to an extreme selection factor right from the start. The same was true for the strong particle stream, which still has a weaker influence on the earth's atmosphere on some days as a solar wind even today. With absolutely no or a very low magnetic field, the particle flow from the sun had unhindered access to everything that developed on the earth's surface. As long as it is not clear when an effective magnetic field existed, the possible particle flow from the sun needs to be discussed as an important parameter.

Another problem is the concentration of molecules supplied by the cosmic particles in the package. Only on a small fraction of the meteorites was it possible for the organic substances to survive on their way through space and the atmosphere that existed back then to the surface of the earth. Smaller particles do not protect the molecules from the UV radiation in space that destroys everything. The next largest units in centimeter and decimeter size are heated up so much when crossing a possible atmosphere that the organic molecules are destroyed. It is no different with really large bodies, which convert so much kinetic energy into heat when they collide with the earth that everything evaporates. Only a small fraction in between is large enough to offer the molecules inside sufficient protection against the radiation in space and not to burn up when they come into contact with the atmosphere or the earth's surface.

The first question that follows this relates to the frequency of events in a catchment area for an assumed pool that dries up. Are the extremely low molecular concentrations (in the range from 1 to 10^{12} per meteorite material) sufficient to start a biochemical evolution with an assumed meteorite event of 1 per year (or per 10 years)? It needs to be taken into account here that the molecules required for the reactions were not all available at the same time. The release took place in the course of the surface weathering and from the smallest cracks, starting from deeper parts, micrometer for micrometer, over a period of hundreds of years. This means that the first organic compounds released were flushed into the flow systems or embedded in the sediment long before the last molecules from the meteorites were available. Only many billions of meteors of the same size weathering at the same time would have been able to provide a resource for the chemical-organic processes required. I have already described above that these molecules were exposed to a large number of destructive mechanisms during their release and subsequent transport. All of the reasoning above is invalid if the surface of the earth was completely iced over before an atmosphere was present (cf. Sect. 2.4).

5.8 Panspermia: Space Seeds

"Space is pulsating with life, unimaginable creatures fight for supremacy in the last galaxy to be defended, spaceships career through space…." Our heads are full of images from all the science fiction films and novels which have thrilled countless people for decades who consider the confines of earth too narrow for expanding our imaginations. The narratives shaped the ideas of entire generations about intelligent, extraterrestrial life shaped by the same weaknesses in character demonstrated by life on earth. Scientists are not free from these thoughts either, although they can assess more precisely the universal laws of physics which science fiction novels appear to override. However, they are not so much concerned with battles fought in space, but much more fundamental things, namely, that of life itself. In the early days of modern science, it seemed very questionable to scientists that life on earth could have originated. In their view, the cosmos offered many more opportunities for the development of life, so that the likelihood of the earth being "impregnated" from the outside was seen as a more plausible alternative. Surprising is the fact that these considerations were discussed more than 100 years ago, at a time when little knowledge of the planets outside our solar system or the laws of space existed. The ideas involved distant planets, on which more favorable conditions existed for the origin of life than on earth. The earth was subsequently infected by a transfer of germs, a concept now known as panspermia. Not only was the transport mechanism unclear, but also their chance of survival in space and once they had landed on earth was speculative.

This early thought, which reminds us of science fiction very much, was even expressed in the 1970s by well-known scientists including Francis Crick and Leslie Organ. Crick, along with James Watson and Maurice Wilkins, received the Nobel Prize in Medicine for discovering the molecular structure of DNA (cf. Sect. 4.2). The chemist Leslie Orgel conducted research in the field of chemical evolution. The two scientists even discussed the possibility of targeted panspermia. According to their ideas, civilizations at risk of extinction on distant planets sent grains containing bacteria into space in an attempt to "infect" distant planets with the germs of life. Doing so would then make colonization possible [16]. This approach can be quickly challenged because the time ranges between vaccinating a planet with biospores and its possible habitability by subsequent intelligent life forms diverge completely. It took over a billion years to produce oxygen on earth alone; it took this amount of time until consumption by oxidative reactions with iron and sulfur had decreased to such an extent that excess oxygen could be enriched in the atmosphere. The civilizations under threat on distant planets certainly did not have this much time available.

Ignoring the reasons given by Crick and Organ, the fascinating idea of vaccinating the earth from the outside raises a lot more questions than it does answer. First of all, in no way does it help to clarify how life came about. It outsources the problem to an unknown region where we have no knowledge of all the general conditions present. Furthermore, there is a lack of information about the conditions in the region from where this launch took place and how the transport could have survived over such very long periods of time under constant bombardment by cosmic rays. Ultimately, the small grains would have any protective cover. The main problem here would be the particle density. From an assumed starting point in space, a huge number of particles would have to be catapulted in a direction that would demonstrate such large scatter after a few light years that actually encountering a planet would be an enormous coincidence. The image of a shower head on the moon sending out a spray of water toward the earth offers a small idea of this. Let us leave out the gravitational pull of the moon and the earth's atmosphere and position exactly one person on the earth. If a jet of water hit the earth at all, the likelihood that precisely this person would be hit would be extremely slim. Or let us take a supernova explosion in a neighboring galaxy that really throws out a lot of material into space: how many particles would arrive on earth? If a particle loaded with bacteria actually managed to strike a planet in the habitable zone and its contents survive the landing approach, the microbes would be confronted with very mundane things such being embedded in sediment or dissolving in aggressive water. The chance of life blossoming would be almost zero.

However, the exchange of matter between two neighboring planets does not seem quite so unrealistic. Billions of tons of rock material have arrived on earth from Mars through the impact of large meteorites. The proportion was much smaller in the opposite direction. However, the amounts may have been sufficient to transport resistant unicellular organisms with the rocks in both directions [17]. The chance is given in theory if the unicellular organisms in the porous rock are covered by a layer of rock at least one meter thick. Even after an expected flight time of more than a million years (on orbits that are slowly approaching the earth), UV radiation is not expected to cause any effect. And that unicellular organisms can survive such a long period of time has been proven, at least in the case of bacterial spores. In this way, it has been possible to obtain at least 25-million-year-old specimens from bees enclosed in amber over this time [18]. The early phase of Mars, which had sufficient water at the beginning, offered quite similar conditions for the development of life as on earth. An interesting consideration is whether life actually came into being twice, perhaps on Mars and earth at the same time. It is difficult to imagine an exchange, however, when the two forms meet. With every genetic code that develops independently, a kind of independent language arises. A biological communication between these two separately developed lines of life would not be possible.

It is well documented in the meantime that organic compounds are formed in space and on other planets. Analyses of the Murchison meteorite that struck Victoria, Australia, in 1969 revealed an astonishing variety of organic molecules which had survived the impact [19]. The presence of 70 amino acids, most of which do not originate from biological processes, and the high percentage of the D configuration instead of the L configuration make contamination unlikely. Even organic bases, which comprise the basic building blocks for RNA and DNA, have been found in carbonaceous meteorites. In some cases, these included bases that do not or only occur very rarely on earth. This is seen as a clear indication of an extraterrestrial origin [20].

All in all, it has been confirmed that the formation of organic chemical molecules is not limited to earth but is also widely possible in space. Explaining the formation of amino acids, lipids, and organic bases in general does not seem to be the problem. The problem is to bring them together in one place so that they can interact in high concentration over a very long period of time and that an engine exists to push these interactions. Only under such conditions is it possible to develop a complex machinery as that of the biological cell.

5.9   Additional Considerations

Numerous other ideas and models exist that deal with individual aspects and provide information about possible detailed special steps in development. A full description of all research methods would quickly go beyond the scope of this book. However, two other methods are worth mentioning briefly. One model also presented in several popular scientific publications includes the process of freezing seawater as the basis for the development of life. The core of the model is made up of a concentration of organic molecules by freezing seawater. On freezing, ice crystals are first formed from freshwater, which causes salts and organic molecules to accumulate in the remaining pores, which in turn are surrounded by ice membranes. This should create favorable conditions for linking the molecules. As has been described for other environments, water represents a major hurdle for linking molecules to peptides or

RNA strands. Freezing removes the pores to a sufficient extent, which means that the reactions can take place. This highly controversial model was developed by the physicist Hauke Trinks from the University of Hamburg-Harburg, who died in 2016 [21]. He assumed that phases of icing existed on infant earth, which cannot be excluded. His model does not explain the origin of the molecules or how the vanishingly small proportions in the primordial sea could bring about the high concentrations needed for reactions in the ice pores.

Another, much-discussed model originates from Manfred Eigen, 1967 Chemistry Nobel Laureate, from the Max Planck Institute for Biophysical Chemistry in Göttingen, Germany. He passed away shortly after my unsuccessful attempt to share my new knowledge and this book with him. Eigen developed mathematical models based on Darwinian evolution, which aim at the self-organization of larger molecules. Self-organization should lead to the formation of self-reproducing units and, at the same time, develop large functional molecules. The terms "quasi-species" and "hypercycle" originate from him [22].

References

1. Oparin AI (1938) The origin of life. Academic Press, New York
2. Miller SL (1953) A production of amino acids under possible primitive earth conditions. Science 117(3046):528–529
3. Müller-Esterl W (2018) Biochemie. Springer, Heidelberg
4. Pedreira-Segade U, Feuillie C, Pelletier M, Michot LJ, Daniel I (2016) Adsorption of nucleotides onto ferromagnesian phyllosilicates: significance for the origin of life. Geochim Cosmochim Acta 176:81–95
5. Wächtershäuser G (1990) Evolution of the first metabolic cycles. Proc Natl Acad Sci U S A 87:200–204
6. Wächtershäuser G (2000) Origin of life: life as we don't know it. Science 289(5483):1307–1308
7. Russell MJ, Hall AJ, Cairns-Smith AG, Braterman PS (1988) Submarine hot springs and the origin of life. Nature 336:117
8. Karson JA, Früh-Green GL, Kelley DS, Williams EA, Yoerger DR, Jakuba M (2006) Detachment shear zone of the Atlantis Massif core complex, Mid-Atlantic Ridge, 30°N. Geochem Geophy Geosys 7(6):21. doi.org/10.1029/2005GC001109
9. Martin W, Russell MJ (2007) On the origin of biochemistry at an alkaline hydrothermal vent. Philos Trans R Soc B 362:1887–1925
10. Kelley DS, Früh-Green GL, Karson JA, Ludwig KA (2007) The lost city hydrothermal field revisited. Oceanography 20(4):90–99, Special Issue on Ocean Exploration

11. Bach W (2011) Weekly report SO-216 (BAMBUS), 15/06-22/06/2011, Townsville, Australia – Eastern Manus Basin, Papua New Guinea. epic.awi.de/37102/20/SO216_wr.pdf (accessed 04/29/2019)

12. Reeves EP et al (2011) Geochemistry of hydrothermal fluids from the PACMANUS, Northeast Pual and Vienna Woods hydrothermal fields, Manus Basin, Papua New Guinea. Geochim Cosmochim Acta 75:1088–1123

13. Darwin C (1887) The life and letters of Charles Darwin, including an autobiographical chapter. John Murray, London

14. Van Kranendonk MJ, Deamer DW, Djokic T (2017) Life on Earth came from a hot volcanic pool, not the sea, new evidence suggests. Sci Am 317:28–35

15. Cnossen I, Sanz-Forcada J, Favata F, Witasse O, Zegers T, Arnold NF (2007) Habitat of early life: solar X-ray and UV radiation at Earth's surface 4–3.5 billion years ago. J Geophys Res 112:1–10. https://doi.org/10.1029/2006JE002784

16. Crick FHC, Orgel LE (1973) Directed Panspermia. Icarus 19:341–346

17. Mileikowsky C, Cucinotta FA, Wilson JW, Gladman B, Horneck G, Lindegren L, Melosh J, Rickman H, Valtonen M, Zheng JQ (2000) Natural transfer of viable microbes in space. 1. From Mars to Earth and Earth to Mars. Icarus 145:391–427

18. Cano RJ, Borucki MK (1995) Revival and identification of bacterial spores in 25- to 40-million-year-old Dominican amber. Science 268(5213):1060–1064

19. Meierhenrich UJ, Munoz Caro GM, Bredehöft JH, Jessberger EK, Thiemann W (2004) Identification of diamino acids in the Murchison meteorite. PNAS 101:9182–9186

20. Callahan MP, Smith KE, Cleaves HJ, Ruzicka J, Stern JC, Glavin DP, House CH, Dworkin JP (2011) Carbonaceous meteorites contain a wide range of extraterrestrial nucleobases. PNAS 108(34):13995–13998

21. Trinks H, Schröder W, Biebricher CK (2005) Ice and the origin of life. Orig Life Evol Biosph 35:429–445

22. Eigen M, Schuster P (1979) The hypercycle – a principle of natural self-organization. Springer, Berlin

6

The RNA World: A Beginning with a Very Special Molecule?

Abstract The RNA world is a term for one of the first steps in development toward life. It lies between the formation of the first organic molecules and the subsequent stage of life in which the DNA took on the function of the information store. The starting point for the discussion on an RNA world was the discovery that RNA can act as both an information store and a catalyst. Problems that arise in the laboratory owing to the double-strand formation of longer RNA molecules do not occur in the crust model because of the larger temperature fluctuations during geyser eruptions.

6.1 RNA, A Molecule with Skills

Geologists like to divide the earth's history into time periods during which conditions were constant over a long period of time. A somewhat fuzzy but more complementary division is made with the help of living beings, which at certain times constituted the predominant groups. The best known is the era of the dinosaurs, which began about 230 million years ago and lasted until the Cretaceous–Paleogene (K–Pg) boundary 66 million years ago. The era of the mammals followed. Prior to the dinosaurs, there were times when, for instance, amphibians, fish, or trilobites existed. There was a suspicion relatively early for the beginning of life that the complex information store DNA represents an evolutionarily higher development than RNA, which can also store information in a way. The considerations originated from the US

© Springer Nature Switzerland AG 2020
U. C. Schreiber, C. Mayer, *The First Cell*, https://doi.org/10.1007/978-3-030-45381-7_6

microbiologist Carl Richard Woese. Building on this, Walter Gilbert, an American biochemist, proposed the term "'RNA world'" in the mid-1980s. The era of RNA was born, the exact beginning and end of which are still unclear. At least since then, many scientists who research the origin of life have been advocating the RNA world hypothesis.

And what makes the RNA world hypothesis so attractive? It was the discovery of a special RNA that is capable of developing catalytic properties [1]. It was found in the ribosome of a eukaryotic unicellular organism, in the molecular tool responsible for the assembly of the proteins. In other words, this RNA can be something that enzymes normally do elsewhere. However, the speed of the catalytic processes is much slower than that of enzymes today. The idea quickly developed that, at the beginning of life, primarily an RNA with catalytic properties existed, and only gradually did the functions switch to the more efficient enzymes. And that is not everything. At the same time, the sequence of the base triplets means that RNA is an information store that only had to be brought into the right relationship with the other molecules. The RNA may even have started the amino acid chaining to peptides, which continues to this day in the ribosomes. In addition, under certain circumstances, the RNA can copy itself, which represents the ideal precondition linking the protein world to the RNA world.

Laboratory tests have shown that ribose, the sugar from RNA, is much easier to obtain from prebiotic starting materials available than deoxyribose, the sugar from DNA. In cells today, deoxyribose is produced with the help of an enzyme from ribose. In addition, the DNA is generated from RNA building blocks, which therefore have to be present first in the cell. These are indications of the life history of molecules, which suggest that RNA existed before DNA. The transition from the RNA to the DNA world apparently had its advantages due to the significantly shorter lifespan of RNA owing to its chemical instability. The reason for this is the slightly different composition of the sugar ribose in RNA compared to the deoxyribose in DNA. While the ribose has an OH molecule at one point in the ring-shaped sugar, at the same point in the DNA sugar there is only one hydrogen molecule that lacks oxygen (deoxy = without oxygen). If OH molecules occur at higher pH values due to the presence of bases, they will extract the hydrogen (H) from the OH molecule in the ribose in order to form a water molecule. This does not happen with the single hydrogen atom in deoxyribose since it is too tightly bound to the sugar ring. With the loss of the hydrogen atom in the ribose, the remaining oxygen immediately combines with the phosphate in the backbone of the RNA. This simultaneously gives up the bond to the next ribose molecule, and the RNA disintegrates [2]. It is hydrolyzed. The reaction also shows that the

RNA is unstable at higher pH values. And from a biochemical point of view, this was exactly one of the arguments against white smokers. Waters with high pH values appear on them, which give RNA no chance of survival. In cells today, the lifespan of the various RNA molecules is limited to a few minutes. Afterward, they have fulfilled their function, are separated into individual components, and become available to assemble new RNA with new tasks. DNA, on the other hand, has a high level of stability with the same storage principle, which is evidenced solely by findings in the bones of people from prehistoric times such as the Neanderthals.

The comparison of RNA and DNA makes it probable that the RNA formed at the beginning of life and, as evolution progressed, the more stable long-term storage DNA developed from it. The RNA was retained and developed into the bearer of a wide variety of functions. But now the question arises as to what the information store looked like or how it functioned before the more stable DNA replaced the RNA.

And so, we touch one of the main difficulties in the RNA world. When DNA took to the stage in the course of evolution, the processes of storing and the information content in the base triplets were very likely already long established. This means that the function the DNA had taken over must have been performed previously by the less stable form of the RNA. Assuming that at the beginning nucleotides had been assembled into an RNA in a favorable environment—without "building instructions," a random sequence of bases resulted with no information content for arranging amino acids in a protein. The matrix that gave the base sequence a logic was missing. The situation is comparable to a long column of the numbers 1–4 in any order without spaces, which fills an entire book. Three of the four numbers should always represent an information unit. Without an interface to a selection system that assigns a precisely defined content to a block of three and at the same time specifies from which of the first three numbers the selection process should start, no information can be selected. However, the strand itself represents information that is passed on each time copying takes place. If a randomly compiled sequence of bases in an RNA receives a certain three-dimensional structure that is, for example, catalytically active, this property is retained by the copying.

An interesting consideration can be made regarding the assignment of the codons from an RNA to a specific amino acid sequence in a chain. Assuming that RNA strands from the very early stages of chemical evolution were found today, then, as with today's RNA strands, three adjacent bases could be defined to form a triplet. The difficulty here is determining a start. The bases in an RNA are all adjacent to one another, which means that the first group of three can begin at three different points. Each time the start is "moved forward" by one base, all the triplets move to a new three-group sort. As a consequence, three completely different assignments exist in the triplet definition for reading an RNA. This does not even matter at first; it only increases the possibilities for subsequent consideration: complementary anticodons of today's tRNAs would certainly be found for the random sequence of bases within each triplet, each specific for one amino acid (with the four exceptions of the start and stop triplets). More than 4 billion years ago, however, today's specifically loadable tRNAs did not yet exist. That means that with the help of modern tRNAs, we could test the RNA from the early days and determine which amino acid chain results from its supposed store. The result is predictable. Completely nonfunctional amino acid chains would emerge each time. In other words, the information content of a random base sequence in an RNA cannot be used as long as a connection to an information system does not exist. This system requires an exact information assignment between RNA, tRNAs, and associated synthetases, which load the tRNAs specifically. At this point we have an egg, but we don't know how it was laid, because we don't have a hen.

6.2   Problems in the RNA World

Previous experiments on the generation of long RNA strands have shown that the degradation proceeds faster than the assembly owing to its own catalytic activity [1, 3]. This means that a randomly formed longer RNA strand has a very short lifespan, since it is broken up into short sections immediately. An interesting discovery in this context was that RNA molecules are linked into longer strands when they circulate in a cell with a thermal gradient [4]. In a laboratory experiment, one side of a capillary a few millimeters thick was heated to over 70 °C while the opposite side was cooled, so that a temperature difference of over 30 °C occurred. While the strands on the warmer side disintegrated, chains were formed on the cooler side [5].

Connecting the concept for the model to existing natural systems is, however, difficult. The conditions in the microchannels in the white smokers were given as an example, from which the higher temperature waters from the oceanic crust reach the seawater. However, the solution load in the water is so high that crystallizing minerals constantly seal the tubules, so that the pathways constantly shift. In addition, the pH values for the solutions that emerge are very high (see Sect. 5.6), so that possible RNA molecules are quickly hydrolyzed.

It is therefore necessary to search for a realistic geological environment for the early days of the earth, which could provide constant conditions over very long periods. Ribozymes (catalytically active RNA molecules) previously developed in experiments in the laboratory demonstrated a relatively high error rate in reproduction and only very short sections could be reproduced [3].

But apart from that—even if a randomly formed RNA copies itself any number of times, suffers errors while copying, and is constantly being enhanced (in the direction of improved catalysis)—it is impossible for it to randomly catalyze enzymes that simultaneously form up to 20 different synthetases, which in turn load tRNAs so specifically that the peptide machinery can develop from them. This would be chicken that hatches from the egg that it laid itself.

References

1. Mills DR, Peterson RL, Spiegelman S (1967) An extracellular Darwinian experiment with a self-duplicating nucleic acid molecule. Proc Natl Acad Sci USA 58:217–224
2. Alberts B, Johnson A, Lewis J, Raff M, Roberts K, Walter P (2002) Molecular biology of the cell. Garland Science, New York
3. Szostak JW (2012) The eightfold path to non-enzymatic RNA replication. Journal of Systems Chemistry 3:2
4. Kreysing M, Keil L, Lanzmich S, Braun D (2015) Heat flux across an open pore enables the continuous replication and selection of oligonucleotides towards increasing length. Nat Chem 7:203–208
5. Agerschou ED, Mast CB, Braun D (2017) Emergence of life from trapped nucleotides? Non-equilibrium behaviour of oligonucleotides in thermal gradients. Synlett 28(1):56–63

7

The New Model: Hydrothermal Systems in the Early Continental Crust

Abstract Fracture zones in the young continental crust form ideal conditions for the emergence of life. In addition to the availability of all the raw materials, and a large variability in pressure, temperature, and pH values, CO_2 gas (gCO_2) occurs in a supercritical phase state. A nonpolar solvent becomes available as a result, in which reactions take place that cannot occur in water. Australia's hydrothermal quartz, which is billions of years old, proves that extensive organic chemistry exists from the earth's early phase in such fault zones. Cyclic pressure fluctuations simulated in the laboratory lead to periodic phase transitions of hydrothermal fluids and the formation of vesicles. At the same time, peptides are being formed which interact with the vesicle membranes and promote structural and chemical evolution.

7.1 The Continental Crust: Fragile and Disturbed

The beginning of the project to address the emergence of life was completely different to any we had known before in our own research. Few scientists had ever set out before to explore life in its early days; few research groups had ever worked on this question in a highly funded research project following the submission of a successful application. In fact, the topic almost stumbled into a situation which had nothing to do with the search for the origin of life. The question that was of interest to me after a certain point in time was as follows: what conditions prevail in tectonic faults? These are fracture zones in the earth's crust up to the mantle, which channel hot water and gases, where minerals are crystallized from dissolved substances and where earthquakes

U. C. Schreiber, C. Mayer, *The First Cell*, https://doi.org/10.1007/978-3-030-45381-7_7

can start. It was the hill-building forest ants repeatedly found on such faults during mapping tours of the Eifel that triggered this question. In addition to leading to known general geological conditions, subsequent fact gathering about gas-open fracture zones led to gases such as nitrogen, carbon dioxide, hydrogen, phosphate, and sulfur and to the conditions found in the Fischer–Tropsch synthesis in hydrothermal systems. And suddenly there it was, an enormous hint that suggested we think about the origin of life...and the rest is history. Following a brief search, a group of natural scientists from the University of Duisburg-Essen was formed as described above, which set itself the goal of transferring today's conditions in tectonic fault zones to the initial phase of continental crust formation and examining the possibilities for chemical reactions with regard to the formation of organic molecules [1].

The hydrothermal systems in black smokers and white smokers have shown that in the transition from the thin oceanic crust to the hydrosphere, a temperature-controlled circulation process creates special conditions for an independent ecosystem. Hydrothermal springs also provide interesting habitats on land for highly adapted bacteria and archaea that tolerate temperatures well above 100 °C. The high temperatures stand in partial connection to deep circulations of water that seep down into higher positions in the mountains and reach several kilometers down into the earth's crust through fracture zones. There they reach the same temperature as the crust and are pushed up again by inflowing water. In the end, the path of ascent is morphologically lower than the mountain region, and the water emerges independently (artesian) as hot springs. In most cases, however, the high temperatures result from the proximity of the water to igneous activities. Magmas exist that get stuck in the earth's crust as they rise in magma chambers and heat up the crust above. Or they can break through the entire crust through fracture zones and form volcanoes. The closer the water is to the hot rock, the more it heats up. Examples of this can be found in Yellowstone National Park in the United States or the Phlegraean Fields in southern Italy with their geysers and hot springs. The temperature gradient in the crust decreases the greater the distance from igneous zones. Today, it reaches average values which amount to approx. a 30 °C increase in temperature per kilometer depth. In the early phase of the development of the continental crust, temperatures may have been twice as high due to higher levels of radioactive isotopes (potassium [^{40}K], uranium, thorium) that made a large contribution to the increase in temperature inside the earth with their decay.

At this point, I would like to emphasize one thing in particular. It is possible to develop and discuss many hypothetical models for the origin of life. However, acceptance can only be achieved if laboratory experiments underpin individual steps in the hypotheses so that a theory can be formed as a result. It is of particular value if general conditions are specified for the proposed model for which realistic physical quantities can be estimated. And that is exactly the case here. We know the pressure in an open water column, which was exactly the same previously as it is today if we leave out atmospheric pressure. It is 1 bar per 10 m. At a depth of 1000 m, 100 bar prevailed at the time even in the young crust, and the temperatures were perhaps 50–60 °C, about twice the value of today. Exactly this information is needed to simulate the conditions of the continental crust in the laboratory.

As described in Sect. 2.5 with the example of Iceland, large amounts of basaltic lavas can be piled up to form islands by being overlaid by various mantle processes, which after a few million years form a large mainland. If parts of these rocks are melted again, new magmas with a completely different composition can form. Slow cooling causes the new molten rock to crystallize into a rock that is related to granite. Granitic rocks have a lower density than basalt and practically float on the denser underground of the mantle. Their continual formation in the course of the earth's history therefore led to the formation of an independent crust, which still forms the core zones of the continents today. The difference in density is the reason why the continents are higher than the oceanic crust and even higher than sea level. And this provides it with a new situation for the early phase of the earth. With the formation of the first small continents, larger areas emerge that are above an assumed sea level. This opens up new spaces that could impact upon life in its beginning phase. Furthermore, the temperature gradient in the continental crust is significantly smaller than in the oceanic crust, while at the same time its thickness is many times larger. This is associated with a much wider pressure and temperature range over the vertical extent, which was available for the development of organic chemical molecules.

Iceland gives us an idea of how the formation of the continental crust may have started. Larger units had already emerged several hundred million years after the formation of the first basaltic crust. As soon as the primary continents reached a critical size, tension must have led to the creation of tectonic fracture zones. The cause of this could, for example, have been local increases in accumulations of magma under the crust at the boundary between the crust and the mantle. This creates a kind of cushion that makes a limited area of the overlying crust bulge. It is stretched, tears open, and forms deep faults from which gases and magmas can rise. A recent example of this is the southern Upper Rhine Graben (Fig. 7.1). In the last 100 million years, an area of about 400 km diameter northwest of the Alps was continuously raised and extensively eroded. The bulging caused it to stretch, causing faults that made

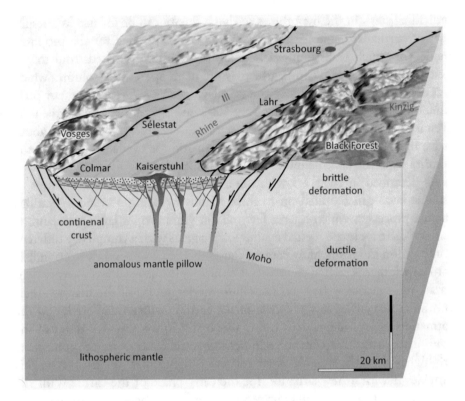

Fig. 7.1 The tectonics of the southern Upper Rhine Graben. The hotter mantle material rising and its accumulation in a mantle pillow led to the entire crust lifting. The stretching associated with this formed the Upper Rhine Rift (© Springer-Verlag GmbH [2], Fig. 13.10)

the Upper Rhine Rift collapse, while the Black Forest and Vosges continued to rise. This development was accompanied by volcanic activity, the best-known representative of which is the Kaiserstuhl near Freiburg. We already have an image in front of us that includes all the preconditions for the formation of organic chemical molecules in the fault zones of a continental crust.

Tectonic Faults

Tectonic faults are caused by fracture zones in the earth's crust where slices of rock rub against one another other. Three basic types can be distinguished between here. The first is caused by the crust expanding. This results in the formation of deposits where the slices sink, as happens, for example, at the edges of trenches. If a constriction occurs, the slices of rocks are pressed against each other, which causes them to be pushed up. When this happens, one floe of rocks is pushed over the other. These two types of faults do not normally form channels at a depth where gases can rise. In large trenches, additional processes overlap that open channels in the deep as in the case of the Kaiserstuhl in the southern Upper Rhine Rift. The third possibility of a fault forming is when the crust tears along lateral displacements by blocks of crust being displaced against one another. By doing so they form vertical fracture zones that can extend into the mantle. The most prominent example is the San Andreas Fault in the west of the United States, where the southwestern part of California is shifting to the northwest compared to the rest of the continent. In the early period, the mantle had a more pronounced dynamic owing to its higher temperature with a presumably stronger convection than today. Stresses that have been transferred to the crust by movements in the mantle caused by shear forces may have been enough for lateral displacements to form. When the crust blocks move against each other, channels open up into the deep on corrugated surfaces that run vertically through the entire crust. Gases, liquids, and magmas can rise at these points.

Faults always occur in a complex network, are interconnected, cross one another, and exchange materials. They also carry water if they are open. Rising gases cause components and water to be transported from a great depth. Water that seeps down into the high mountains can penetrate deep into the crust and rise artesically through other fault paths. From the very beginning, a substantial proportion of atmospheric gases were emitted from the mantle through fault zones of this type, which also occur in a similar form in the oceanic crust. The starting substances for organic chemistry can be derived from them. This means that the raw materials for biological development were available in unlimited quantities if we go by the tectonic fault model. These include carbon dioxide, carbon monoxide, nitrogen, hydrogen, ammonia, and even sulfur compounds. A phosphate is also included that acidic water could extract from the mineral apatite which is a calcium phosphate.

Clear evidence of previously open areas in fault zones, for example, are provided by ore veins, which have been closed by metal sulfides crystallizing, and minerals veins, such as quartz and calcite (Fig. 7.2). The repeated opening and supply of ore solutions led to characteristic changes in ore and minerals veins, which crystallized in a mirror image on both side walls that formed a boundary. These veins, which can contain a wide variety of metal sulfides such as copper, iron, or zinc sulfide, must have been formed in water-bearing fault zones from the start in the continental crust. They document the wide range of metallic sulfides that lined the boundary surfaces of the faults as wallpaper. Pyrite and other metal sulfide surfaces can be found here, which for Wächtershäuser played a role in the metabolism (see Sect. 5.4).

And now we come to the characteristics of an open fault zone that is of outstanding importance for prebiotic processes. With an adequate supply of the cocktail of gases we are already familiar with, it represents a highly productive chemical factory for the production of organic molecules (Fig. 7.3).

Are the conditions that prevail in a fault something new? Only partially. In technical chemistry, long-chain organic molecules are produced from carbon monoxide and hydrogen with the help of the Fischer–Tropsch synthesis under

Fig. 7.2 Ore vein with metal sulfides (sphalerite, chalcopyrite), calcite, and some quartz, Bad Grund/Harz

high pressures and temperatures (e.g., for gasoline). At temperatures up to 300 °C and pressures of up to 25 bar (which corresponds to a water depth of 240 m at 1 bar air pressure), this gives access to the formation of, for instance, alcohols or methane, propane, and even more complex compounds. Further reaction steps can be derived from this to produce a large number of hydrocarbon chains that are required for biochemical development, including molecular chains that are required for building a cell membrane. Various metallic catalysts or the mineral zeolites ensure a high yield in the reaction.

And that is exactly what we had in mind with our continental fault model. From the very beginning, a multitude of pressure and temperature conditions existed in the open water column that go far beyond that which is specified as the framework for the Fischer–Tropsch synthesis. In addition, the supply of starting materials and metallic or mineral catalysts was and still is much

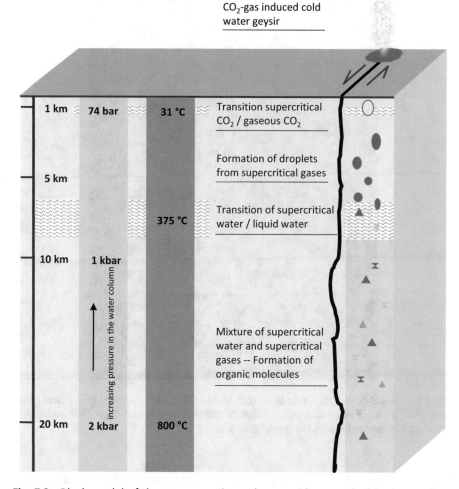

Fig. 7.3 Block model of the young continental crust with up to double the geothermal gradient as compared to today (30 °C per km today). On the right, a view of the fluid phases in a vertical fault zone. The colored symbols symbolize organic molecules formed locally

higher, as the example with the veins of ore shows. The boundary surfaces in the fracture zones are rough, crack frequently, and contain a large number of small protrusions and cavities in which rising gases can collect (Fig. 7.4). Strictly speaking, small micro-autoclaves exist in infinite numbers, whereby each autoclave, which is a kind of pressure cooker, has its own special set of physicochemical conditions (Fig. 7.5). This gives you an idea of the potential

Fig. 7.4 Typical course of a tectonic fault. When filled with water, supercritical gases can collect in the cavities at depths of 1000 m (arrows). In this case, they form autoclave-like reaction spaces (Photo: Dr. Frederik Kirst)

that lies in such a fault system. A reaction vessel exists for almost every one of the most important organic compounds required in the course of originating life, which provides the right pressure and the right temperature for its synthesis. Even the pH value can be varied within certain ranges through the mixing ratios for the gas components CO_2 and N_2. While a high CO_2 share under pressure results in pH values in the acidic range up to a minimum of 3.3, the pH value can rise above 6 by increasing the nitrogen content. If higher concentrations of sulfur compounds are involved, the pH can fall below than 3.

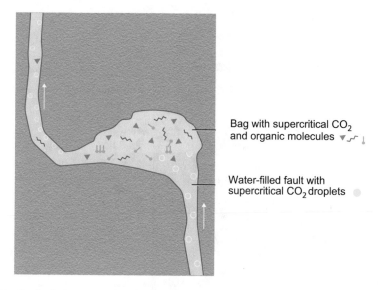

Bag with supercritical CO_2 and organic molecules

Water-filled fault with supercritical CO_2 droplets

Fig. 7.5 Cavity (micro-autoclave) in the continental crust with water, supercritical (sc) CO_2, and organic molecules

7.2 Supercritical Gases: Steam Under Pressure?

The life of a researching natural scientist may have one, for a few several, for many absolutely no personal magic moments in science. I have defined one for myself, an almost banal snippet of knowledge, retrieved from the Internet at a time when enthusiasm for the question of the origin of life in our group has already given way to a certain amount of disillusionment.

Gases assume a special phase state from a specific temperature and pressure. They become supercritical. The density of supercritical gases is about half that of the same substance in liquid form. With CO_2, the transition from gas to the supercritical phase takes place at around 31 °C and 74 bar.

And then suddenly I came across it by chance in the Internet: CO_2 can already become supercritical at an elevated temperature in the earth's crust at a depth of 740 m. I'll never forget that moment late in the evening at 11 p.m. Applied to an open fault in the crust containing a CO_2-saturated water column, this meant that the transition to supercritical CO_2 for early earth must have taken place from a depth of about 740 m (Fig. 7.6). The prevailing temperature gradient at the time was certainly enough for this. Today, the transition at the average temperature increase of 30 °C/km usually takes place at

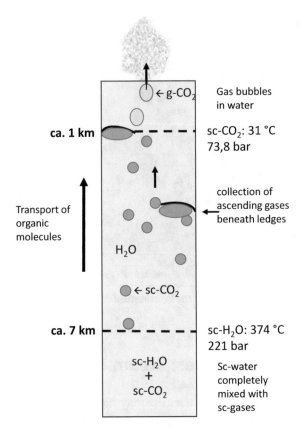

Fig. 7.6 Schematic diagram of a water column in a gas-permeable fault zone with a geothermal gradient twice as large as today. Below 7 km, a fluid made of supercritical (sc) water, CO_2, and N_2 (not shown) exists. Over and above this, the supercritical gases are separated from the now liquid water in the form of bubbles. From about 1000 m to the surface, gaseous carbon dioxide (g-CO_2) exists next to water

the 1000 m mark. In volcanically active regions, the temperature of 31 °C is already reached at a lower depth. The pressure then becomes decisive for the phase change.

In addition to CO_2, nitrogen (N_2) was also present as a gas in the crust in the early phase of the earth's development. Like any gas, N_2 can also become supercritical. This happens above a critical temperature of −147 °C and a critical pressure of 33.9 bar. Pure nitrogen gas in a saturated water column would therefore enter the supercritical phase at a depth of around 340 m. Supercritical gases can be mixed with one another at any ratio. Depending on the concentrations of the substances involved, the mixtures have other critical values that deviate from the individual values for the pure gases.

In its supercritical state, CO_2 behaves like an organic solvent which also has a very low surface tension. It can dissolve nonpolar organic substances which are not soluble in water. This property is used for an approach which is known as "green chemistry." A common example of this is the decaffeination of coffee beans. They are rinsed with supercritical CO_2 under pressure which represents a nonpolar solvent capable of extracting caffeine. The pressure is then reduced until the supercritical CO_2 transforms into its gas state, causing the caffeine dissolved in CO_2 to precipitate. It can then be collected and used for other purposes. In contrast to liquid organic solvents, no potentially toxic residues remain in the coffee beans. Together with water, the supercritical CO_2 forms a coexistence of two solvents which is known as a two-phase system and which facilitates a very special class of chemical reactions.

This was a very new aspect: a model for the origin of life in which a nonpolar solvent played a role. This model suddenly allowed reactions to be discussed that could not take place in water but which were required for many of the steps on the path to life. That meant we could take another look at the fault zones in the first continental crust with different eyes. We now could propose the following scenario: geological processes caused large quantities of CO_2 to get into the earth's mantle. During the cooling process, large volumes of CO_2 originally dissolved in the magma were released. The reason for this was (and still is) the crystallization of typical shell minerals such as olivine or pyroxene due to cooling which are not capable of absorbing CO_2 into their crystal lattice. The density of the supercritical gases is significantly lower than that of the mantle, which consists of a mixture of minerals and molten rock. The CO_2 forms droplets and streaks which, when they reach a sufficient amount, rise through the mantle. Depending on their occurrence, they mix with supercritical water and other gases and accumulate in the uppermost regions of the mantle. In the early days of the earth, their ascent stopped at the interface with the continental crust, which formed a barrier. However, if fault zones had formed that penetrated the entire crust right down to the upper mantle, then the CO_2 could rise to the surface through the channels in the fault (this process still takes place today with weakened intensity).

The whole thing is, however, a little more complex than I have described so far. But all in all, the properties of the supercritical phase are extremely supportive for complex prebiotic chemistry. It is simply impressive to see the possibilities that the fault zone system harbors in the crust. As I explained above, the supercritical phases, i.e., the gases described, mix indefinitely, as do the subcritical gases. One example is air, basically a mixture of oxygen and nitrogen.

An additional fluid component inside fault zones is water. Like almost all substances, it occurs in the phase states solid, liquid, and gaseous, i.e., ice, water, and steam. And of course it has a supercritical phase, which, based on the pressure and temperature conditions, corresponds to the conditions of in lower earth's crust, more precisely, from 374.12 °C and 221 bar (2210 m open water column). That means that today, depending on the temperature of the crust, water in its liquid form only appears above a depth of around 13 km in an open water column. The boundary was higher for the hotter crust on infant earth, at perhaps 8 km crust depth. Below that, water was supercritical and mixed completely with the gases which were also supercritical. Now it is worth taking a closer look at the boundary between supercritical water and subcritical water. While the water becomes liquid in the transition caused by decreasing temperature, the gases remain supercritical and separate. This means that they form bubbles of supercritical gas that rise in the water column. This takes place because their density is lower than the now liquid water. As soon as the water column is saturated with CO_2, the droplets make it to the top without being released. It's like opening a bottle of a carbonated beverage, in which bubbles immediately form and move to the top.

With this in mind, a far-reaching conclusion comes into view. We are already familiar with the many small reaction chambers, our micro-autoclaves, in which the most diverse range of organic molecules can be formed over the crust's entire vertical extension. And with supercritical CO_2, we have a non-polar solvent that absorbs all substances that do not dissolve in water. These therefore represent ideal conditions for the supercritical gas droplets. The droplets move through the entire crust and collect specific molecules on the way. At the same time, they create a gentle flow of water through which other molecules, which are dissolved in the water, are transported. If on their journey upward, the CO_2 droplets collect in cavities and protrusions in the fault zone, small reaction chambers are formed, which are separated into two different phases. Supercritical CO_2 remains suspended in the head space while underneath still or lightly flowing water can be found. Besides water-soluble organic molecules, the water contains different concentrations of dissolved salts, which cannot be dissolved in the supercritical CO_2.

And now one of the most important potentials of the system becomes clear: in the supercritical CO_2 and especially at the interface to the water, reactions are possible that would not take place in the water alone. There, water-soluble substances can react with water-insoluble substances, thus forming amphiphilic soap-like components that can create membranes. These conditions also favor reactions in which a water molecule is released during the linking (condensation reactions), e.g., in the reaction of amino acids with each other to form peptides, the amino acid chains.

7.2.1   The Decisive Step

The next step on the way up takes place at a depth of between 1000 and 750 m, which for us gradually turned out to be a possibly decisive step in the development of life. Depending on the temperature and density of the topmost water column, which depends on the number of gas bubbles rising, the supercritical CO_2 droplets become subcritical. That means that CO_2 gas is created that rises through the water in the form of bubbles. The now gaseous state of CO_2 prevents organic components from being kept in solution, just as described in the example with caffeine. In addition, water originally dissolved in supercritical carbon dioxide now condenses in small droplets. Precipitating organic substances now concentrate in those water droplets and in the remaining aqueous solution. Amphiphilic soap-like products preferably accumulate at the water-gas interface, which is now available in the small autoclaves slightly above the boundary zone.

In the early days of the earth, as well as in part today, a depth range existed below around 800 m in which a constant change from supercritical to subcritical gas took place in the micro-autoclaves. The reasons for this were the cyclical pressure fluctuations, which occurred daily owing to earth tides or eruptions of gas in cold water geysers. At the time when life began to develop, the moon was much closer to the earth than it is today. The effects of this were considerable. Huge tidal waves developed in the oceans, which flooded large areas of the first mainland areas. But not only masses of water are attracted to the moon. The solid crust of the earth cannot resist its attraction either. Even today regions exist where locations can be raised and lowered twice a day by up to 40 cm. As with the water masses, the force of attraction on the crust was much stronger at the beginning. This resulted in cyclical pressure fluctuations, which caused cyclic phase transitions between the supercritical and the subcritical state in the boundary zone. A recent example of this can be found in Wallenborn in the volcanic region of the West Eifel in Germany, a CO_2-driven cold water geyser that erupts approximately every 30 to 35 min (Fig. 7.7). Depending on the intensity of the eruption, which catapults a great deal of water up to a height of four meters, the pressure column is relieved at depth, primarily by the water catapulted out and secondly by the high number of gas bubbles in the water column above the boundary zone, which reduce the density of the water-gas mixture. If the pressure decreases at depth, the supercritical state can no longer be maintained in the last caverns below the boundary. Turbulent degassing takes place, which continues at depth until the limiting pressure for the supercritical state is no longer undershot.

Fig. 7.7 Wallenborn cold water geyser, West Eifel, Germany. CO_2-driven discharge of water up to a height of 4 m every 30 to 35 min

The change between the two states can take place in a section of several hundred meters. The water ejected then flows back, whereby the pressure builds up again.

This can recognize a new important step here as a result that immediately solves a problem in the discussion on the origin of life: the high concentration of organic molecules required. They are transported up from the depths, sometimes react with one another on the way up, and are discarded on reaching a certain boundary zone. A space immediately opens up in our minds here, which we can associate right away with an infinite number of reactions. It is obvious for us to assume that these represent the starting conditions for the formation of the first cell. Today, based on these new findings, tests can be carried out that are based on generally applicable physicochemical laws and on realistic parameters. The exciting thing about this is that even today, processes of a similar kind still occur at depth in carbon dioxide sources. We just do not get to see the results of this however, because microbiological activities cover up all the information about it. But we will address a few more thoughts to this in the last chapter, however.

If all the factors in favor of biogenesis are collected and weighted, our enthusiasm for the continental crust fault zone model can be understood. Through contact with the earth's mantle, the fault zones offer ideal conditions for organic chemical reactions. Due to the continuous outgassing of the earth and the decomposition of minerals, all the raw materials required are available in large quantities and over very long periods of time. They can react at different depths with different pressure and temperature conditions and pH values to form larger molecules, be transported along with rising fluids, and also be concentrated in a narrow zone. Experiments by colleagues from different countries have already shown that the key building blocks for life, such as lipids, amino acids, and organic bases, can form under hydrothermal conditions [3]. Another advantage is that the conditions remain stable over very long periods of time, i.e., many millions of years, that destructive UV radiation or plasma particles from the solar wind do not strike, and that the impact of meteorites has little influence—conditions that we can search for in vain on the face of the earth. Rising supercritical gases and cyclical pressure fluctuations can also generate energy and entropy gains that are the driving force behind the early development of life. It is obvious that a knowledge of these relatively well-defined general conditions means that experiments can be carried out which give us access to the individual steps in the development of life.

7.3   It Does Exist: Evidence from Nature

Another flash of inspiration shot through the evening round. Perhaps it was the spaghetti in the overfilled bowl I passed on to my neighbor after I had received the first serving that was the precursor to this. One of our parties had cooked it. I'd arrived at the meeting a little earlier and had just entered the kitchen when a colleague pulled a not inconsiderable mound of pasta from the sink, before rinsing it briefly under water and plonking it on top of the main pile in the colander. As is not uncommon when cooking spaghetti, some of it had been forced over the edge while dousing it with water.

Something that has been forced out of a deep channel must contain traces of what can be found below. So, what about the continental crust? If we already had clear ideas about the conditions that prevail there, shouldn't nature also be able to provide us with evidence that confirmed our considerations? It quickly became clear that this might be documented in one form or other. You only had to find the right rocks or minerals of a correspondingly high age that originated from the faults in question.

In the fault zones in the earth's crust, various minerals crystallize from hydrothermal solutions, such as quartz, which form well-formed rock crystals with enough room to grow. If the crystals are milky, they contain liquid inclusions which consist of a water–gas mixture without any subsequent overprint, as was present in the fault zone at the time of growth. This description makes it clear that "frozen" documents about the chemistry of the aqueous components in hydrothermal quartz must exist, which date from the time when they crystallized. And if organic chemistry had occurred in the liquids, we should be able to determine it. The older the crystals are, the closer you get to the time when no biological activity existed on the face of the earth and the better the chance of identifying the primarily inorganic chemistry that led to organic products.

Didn't that present a great opportunity? If we could find organic chemistry in liquid inclusions in old hydrothermally formed quartz, it would be a sensation. Not only would we get evidence of how billions of years old organic chemistry was composed, but also the findings could support the hypothesis that life began in the top crust of the earth.

These considerations led to a spontaneous decision. It had to be possible to find quartz minerals from the cores of ancient continents that had crystallized in hydrothermal fissures billions of years ago and which have retained the composition of the water in their inclusions to this day. The search for accessible regions with pre-Cambrian quartz vein deposits finally led us to Western Australia, to the Jack Hills region, 900 km north of Perth. Here, hard quartz veins course right through the landscape like the crumbled remains of a wall (see Fig. 7.8). They are younger than 2 billion years old, so that a contamination of the hydrothermal waters by biological material, which has existed for more than 3 billion years, could not be excluded. Nevertheless, we sampled them during two field surveys to clarify the basic question of preserving organic matter in liquid inclusions. We also found quartz pebbles from sedimentary rock (conglomerate) that occurs in a 2.7-billion-year-old sequence of layers in Jack Hills in Western Australia of particular interest (Fig. 7.9). The oldest zircon minerals on the face of the earth dated to nearly 4 billion years have already been found in this sedimentary rock [4]. Like all other components in a sediment, the zircons were created elsewhere, later exposed by erosion, and transported to the deposit site, as were the individual quartz rocks

Fig. 7.8 Quartz dyke in Western Australia

in the conglomerate, whose age cannot be determined, however. They can come from residual solutions that originate during the crystallization of granitic magmas, which then makes them of no interest, or from hydrothermal crevice systems in the crust. With the quartz crystals from Jack Hills being identified as hydrothermal, we had a chance that, besides the scree that was a little more than 3 billion years old, we would also encounter some scree that had crystallized in the crevices of the first continents even before the appearance of LUCA, perhaps more than 4 billion years ago, and was only exposed later, carried away, and smoothed during transport.

The analyses of the liquid inclusions in the hydrothermal quartz were carried out by the research groups headed by Oliver Schmitz (Applied Analytical Chemistry, University of Duisburg-Essen), Heinfried Schöler, and Frank

Fig. 7.9 Conglomerate from Jack Hills, Western Australia

Keppler (both Heidelberg University) [5]. They brought a surprise, a result which we had not expected with such clarity. Data from 1 of the 20 quartz veins and pebbles sampled from Jack Hills provided evidence of the presence of rich organic chemistry in the early hydrothermal systems of the earth's crust. The inclusions contained long-chain and smaller organic molecules, which occur in a similar form in the metabolism of a living cell. In order to rule out subsequent contamination with biological molecules by early waters, methane, as the main representative of the organic compounds, was examined for its isotope composition. In this case, for instance, different carbon isotopes exist, which are stable and do not decay. These are the ^{13}C and ^{12}C isotopes that occur in a certain ratio depending on the carbon source. Through metabolic processes that tend to use the lighter isotope, the ratio shifts accordingly to lighter values while the carbon is part of the biosphere. This allows organic molecules originating from biological processes to be recognized. The isotope ratios of the methane analyzed from the samples examined showed that it comes from an abiotic source.

Perhaps it was just a coincidence that the samples taken in Jack Hills show exactly the conditions for a hydrothermal fault environment with a rich supply of organic molecules. Or maybe most of the pebbles are already equipped with a comparable chemistry. Further sampling and analysis will show whether much more information can be obtained from the quartz. The fluid inclusions have an astonishingly diverse composition, which in some cases continues to raise clear questions. The results of their analysis are of great value to continuing considerations and particularly for experiments in the laboratory. The results provide information about real compositions that allow focused experiments to be performed under realistic conditions. They make certain steps in the model verifiable, a possibility that was not conceivable in the previous model.

7.4 Experiments on the Crust Model Are Possible

The time was ripe for experimental approaches. Any hypothetical model remains just that until detailed steps can be completed by performing plausible experiments. The discussions about our hypothetical model were all nice and good and also pretty advanced, but without hard data from experiments, we sooner or later got the feeling that we had come to a dead end. Experimental equipment capable of simulating the conditions in the upper crust is commercially available.

The high-pressure system made it possible for us to finally carry out tests using real parameters and substances that appeared realistic to us. The information needed for the experiments about the general conditions could be taken directly from earth's crust because of the conditions that still exist there today. This made it possible to subject water and gases to varying pressure and temperature conditions that corresponded to natural conditions. For the reactions, specific molecules were added that we were already familiar with from hydrothermal systems. The analytical results from the liquid inclusions in the Australian crystals proved extremely valuable for the latter.

7.4.1 About the Device

The system can be used to simulate conditions in a reaction chamber, like those that occur in the top 10 km of the continental crust (Fig. 7.10, High-pressure device from SITEC-Sieber Engineering AG, Switzerland). The chamber has a total volume of 50 ml. For the experiments, half of it is filled with water and the other half with CO_2 which is compressed to a supercritical state. Organic molecules, such as fatty acids, amino acids, or RNA building blocks, are added depending on the question posed.

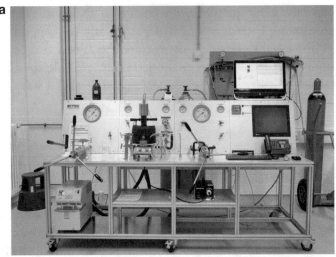

Fig. 7.10 (**a**) High-pressure system (SITEC-Sieber Engineering AG, Switzerland) pressure buildup over right spindle, isolated pressure chamber in black. Pressure fluctuations can be applied with any length of intervals, which simulate natural geyser eruptions. (**b**) Pressure chamber with glass window toward the reaction space (Photos: Yildiz Danisan)

The first experiments designed by Christian Mayer using the device led to a key experiment succeeding in a breakthrough for our further considerations. A previously unanswered question addressed the formation of vesicles under prebiotic conditions, which ultimately forms the basis for a cell. We had specific ideas about how vesicles could form in the fault zone. Simulating the conditions in the high-pressure chamber was simple. A mixture of long-chain amines and fatty acids, as can be formed from the building blocks found in the liquid inclusions in hydrothermal quartz, was transferred to the reaction chamber and kept under the prevailing temperature and pressure in the upper crust for a period of 24 h. When a controlled pressure loss took place, the previously supercritical CO_2 changed into the gaseous state (Fig. 7.11a steps 1 and 2).

A small percentage of water is always dissolved in supercritical CO_2 ($scCO_2$). This water has exactly the same problems as the also dissolved organic substances when the $scCO_2$ changes to the gas phase during the pressure drop. It has to precipitate in some form, because at this concentration,

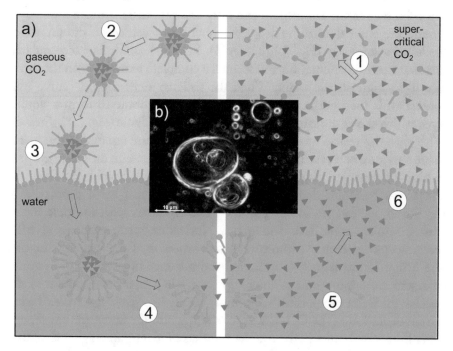

Fig. 7.11 Vesicle formation in the border area between supercritical CO_2 and CO_2 gas (gCO_2) [4]. (a) Cycle of the phase transitions, starting with supercritical CO_2 in a cavity (1). (b) Microscopic image (dark field) of vesicles, several of which are of the multilamellar type (Photo: Dr. Maria Davila Garvin)

it cannot be absorbed in the newly formed gas phase. This results in condensation in small droplets as the pressure drops, most of the drops measuring only a few micrometers.

As a consequence, fog-like phenomenon formed during the experiment, which was clearly visible through a window in the now gas-filled section of the autoclave. Apparently, the droplets are the medium that the organic molecules spontaneously recognize as a refuge. Although, under other conditions, they prefer to stay away from the water, the phase change leaves them with no choice but to collect either in the droplets or on their surface. This leads to the water droplets being highly loaded with organic molecules. Salts, such as those normally found dissolved in the water in the fault zone, are not found here. In this regard, the droplets correspond to distilled water.

7.4.2   So, What Happened Next?

A step followed that brought us a great deal closer to understanding how a vesicle, a precursor to a cell, could have formed under the conditions prevalent on infant earth. The amines and fatty acids (lipids) formed a monolayer around the outer skin of the droplets which slowly settled on the surface of the lower water body (Fig. 7.11a, Step 3). Here too, lipids had accumulated in a characteristic orientation and formed a complete monolayer on the interface like a film on the water. When the sinking water droplets come into contact with the interface, the two monolayers combine to form a double layer, which shows the same structure like a cell membrane (Fig. 7.11, Step 4). With this, the vesicle was completed, and its structure resembled the membrane structure of a cell. It consisted of distilled water with an increased proportion of organic compounds on the inside and a lipid bilayer on the outside—a structure that can be considered the basis for a protocell. In Fig. 7.11b, we can see some "normal" vesicles, others with multiple membranes, in microscopic dark-field images. Nuclear magnetic resonance (NMR) spectroscopy [6] clearly demonstrates the existence of the vesicle structure. The first measurements also indicated the presence of concentration gradients, which are important as an energy source in a later development of the vesicles: the water droplets collect a whole variety of organic molecules during the condensation in the CO_2 gas. After sinking into the water, the concentration of these molecules in the droplet is orders of magnitude higher than that in the surrounding water. On the other hand, no salts are dissolved in the condensed water droplets. Transferred to the salt-rich host water in hydrothermal fault zones, this means that the vesicle formation process builds up two opposite concentration gradients. A first protocell could have drawn the energy for a simple metabolism from these gradients. Steps 5 and 6 in Fig. 7.11a indicate the breakdown of the vesicles and the new beginning of the lifecycle. One cycle consists of the buildup of pressure up to the formation of supercritical CO_2 and a subsequent relief of pressure, so that the gas phase is generated again.

Christian Mayer devised further experiments which again showed amazing results. The first question asked whether the conditions in the pressure cell are suitable for linking amino acids into chains—something that under normal conditions actually requires enzymes. For this purpose, 12 different amino acids, which are known to be formed in hydrothermal systems, were added to the chamber (Fig. 7.12). The result was again very surprising. Within days to weeks, the amino acids combined into peptides with lengths of up to 18 units. This was quite encouraging. Of particular interest was the question of whether there is a possible preference for certain peptides in interaction with the structure of the vesicle membrane. An initial long-term run of the experimental setup gave clear indications that a mutual influence between vesicles and peptides exists under the given conditions, resulting in a selection of certain amino acid chains. This would represent an unprecedented experimental proof of a chemical evolution of peptides under realistic conditions. In the

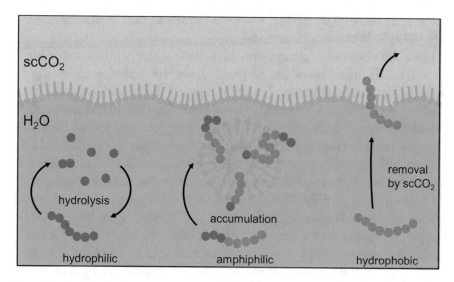

Fig. 7.12 Selection of peptides: 12 amino acids present in proteins are formed hydrothermally: 6 are nonpolar (orange) and 6 are polar (blue). Chains formed by polar ones prefer to be in water (left); chains formed by nonpolar ones are released into the supercritical CO_2 (right). Amphiphilic chains can interact with membranes and integrate into the vesicles (center)

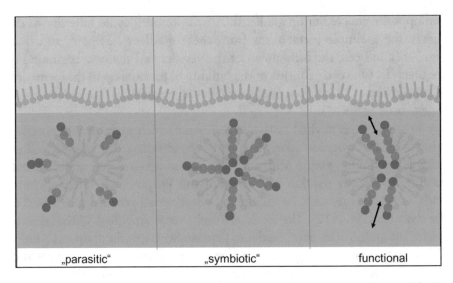

| „parasitic" | „symbiotic" | functional |

Fig. 7.13 Development of peptides in connection with vesicle membranes [8]. The "parasitic" ones only use the protection of the membranes, the "symbiotic" ones are protected but at the same time stabilize the vesicle, and the functional peptides, e.g., form a channel, which enables equilibration of concentrations, preventing damage due to osmotic pressure

meantime, larger peptide molecules have been identified which alter the vesicle's size, increase their stability, and form channels in the vesicle membranes (Fig. 7.13) [7, 8].

From the test results, certain dependencies that the peptides formed have on the vesicle membranes can be seen. There are amino acids which are attracted to water (hydrophilic) and repelled by water (hydrophobic). Interesting chains can be combined from them. Peptides that consist only of hydrophilic amino acids will remain in the water alone and will not seek contact with vesicles. Hydrophobic peptides are washed out by supercritical CO_2 droplets and are permanently transferred into the $scCO_2$ phase. However, chains combining hydrophilic and hydrophobic amino acids also exist which are referred to as amphiphilic. For them, the vesicle membrane, which consists of two layers, is interesting. The outer shell is made up of long molecules that point outward with a hydrophilic head, while the long tail part is hydrophobic and points inward. The inner membrane is built using the same molecules in exactly the opposite direction. The hydrophilic head points inward to the enclosed water, and the tail part points outward or into the middle of the membrane. This creates a hydrophobic environment ideal for those (hydrophobic) amino acids that have nothing to do with water. If peptides are present that have hydrophilic amino acids on one side and hydrophobic amino acids on the other, they can try to fit into the membrane structure such that each amino acid is surrounded by an environment of suitable hydrophilicity. You can now play through many combinations, one of which is particularly interesting. Peptides exist that have a hydrophilic section on both sides and a hydrophobic middle section. If they are integrated into the vesicle, they form anchors that stabilize the cell. With favorable constellations between adjacent amino acids, small channels are created which allow for the molecules' migration of small molecules (Fig. 7.13). This results in an equilibration process that could even be used for generating energy.

References

1. Schreiber U, Locker-Grütjen O, Mayer C (2012) Hypothesis: origin of life in deep-reaching tectonic faults. Prebiotic Chem Orig Life Evol Biosph 42(1):47–54
2. Meschede M, Warr LN (2019) The geology of Germany. Regional geology reviews. Springer, Berlin
3. Fujioka K, Futamura Y, Shiohara T, Hoshino A, Kanaya F, Manome Y, Yamamoto K (2009) Amino acid synthesis in a supercritical carbon dioxide-water mixture. Int J Mol Sci 10:2722–2732
4. Wilde SA, Valley JW, Peck WH, Graham CM (2001) Evidence from detrital zircons for the existence of continental crust and oceans on the Earth 4.4 Ga ago. Nature 409:175–178
5. Schreiber U, Mayer C, Schmitz OJ, Rosendahl P, Bronja A, Greule M, Keppler F, Mulder I, Sattler T, Schöler HF (2017) Organic compounds in fluid inclusions of Archean quartz – analogues of prebiotic chemistry on early Earth. PLoS One 12(6):e0177570. https://doi.org/10.1371/journal.pone.0177570
6. Mayer C, Schreiber U, Dávila MJ (2015) Periodic vesicle formation in tectonic fault zones – an ideal environment for molecular evolution. Orig Life Evol Biosph 45(1–2):139–148
7. Mayer C, Schreiber U, Dávila MJ (2017) Selection of prebiotic molecules in amphiphilic environments. Life 7:3. https://doi.org/10.3390/life7010003
8. Mayer C, Schreiber U, Dávila MJ, Schmitz OJ, Bronja A, Meyer M, Klein J, Meckelmann SW (2018) Molecular evolution in a peptide-vesicle system. Life 8:16. doi.org/10.3390/life8020016

8

A Hypothetical Approach: Hydrothermal Systems in the Early Continental Crust

Abstract The development of life can be divided into six phases. Organic molecules can be formed vertically over long distances in the continental crust. Rock minerals' dissolving releases phosphate, metals, boron, and other substances. A process of collection caused by rising supercritical gases leads to a strong accumulation in cavities (micro-autoclaves) and in the transition zone to the subcritical gas at approx. 1000 m. This is the place where vesicles and peptides are formed. The conditions are optimal for the energy required and the increase in entropy. The requirements for the formation of the first cell are discussed in a hypothetical model. According to the model, following the formation of the first unspecific synthetases and tRNAs, a selection of proteins began to form out of only two amino acids. Simultaneous storage of the information in an RNA enabled the development of specific synthetases with which the principle of life started.

8.1 The Search for the Path

The start was successful. Our first experiments had shown that we were on the right track. Our understanding of the environment in which the possible origin of life existed was getting better and better. The experiments under clearly definable conditions provided the first indications of realistic scenarios, and additional calculations of physicochemical processes by Christian Mayer offered a secure basis for further experimental setups. The presentation of the

© Springer Nature Switzerland AG 2020
U. C. Schreiber, C. Mayer, *The First Cell*, https://doi.org/10.1007/978-3-030-45381-7_8

results at numerous international conferences has met with great interest. What was missing, however, was the integration into an overall development, which should ultimately result in information storage and the formation of the first cell. In other words, a concept was missing, an idea of how things might have gone overall. Didn't a path exist to the solution that could be pointed out to us, taking into account our own and international research results on the origins of life, for which no restriction in the combination existed initially, where everything was permitted, provided that the fundamental physical and chemical laws were observed?

To make it clear once again, it is a problem in science to set up hypotheses for processes whose causes experiments cannot fully prove. However, research into the origin of life presents a special case. Too much time has passed since the first steps took place, and the conditions on infant earth are simply too obscure to reconstruct them with just a few experiments. Experiments without any hypothetical superstructure, however, inevitably run into a dead end. The connection to the whole is lacking.

We had arrived at a point where it should have been. We needed one or more hypotheses in which we could play through the possibilities for how life started in different variations. The formulation of a hypothesis relating to this field of research is similar to an attempt to convict a perp in a flimsy trial underpinned by circumstantial evidence. The experiment can easily be misleading. In relation to the processes on infant earth, some indications have become apparent in connection with the new crust model, which provide a framework with the first reliable data. This made it logical therefore to test a hypothesis that describes the development from the simplest formation of an organic molecule to a dividing cell for precisely this environment. Such a comprehensive hypothesis inevitably raises many questions. However, questions are a prerequisite for developing focused experiments that can provide answers in the search for a solution. The future will show whether the experiments can be used to find supportive information for the model or whether established assumptions must be rejected on the contrary.

Something in the discussion about the origin of life had rattled me from the very beginning. It was a gut feeling that became stronger the more we progressed with our experiments. On the one hand, we had succeeded in developing peptides relatively easily in our experiments, which may mature up to the size of enzymes with more time perhaps, but on the other, no possibility existed for storing the information about their structure. Each large molecule had its own amino acid sequence initially. After the decay, the probability of an identical molecule forming with these gigantic variations in possibilities was practically zero. Only similar molecules could be formed, which at best could be divided into groups. The representatives of the RNA world take a different approach and start immediately with an RNA molecule. However, they do not explain how, at the beginning, the base groups in the first RNA strands could be defined as information units, thus making possible an exact assignment of amino acids in the enzymes. Enzymes and RNA were needed from the start—and here we have our famous chicken and egg problem. And with every consideration, the infinitely large number of possible variations resounded in the back of your minds. What shape could a solution to all our problems and questions take?

It took many long weekends making stacks of notes covered with symbols, chains, and flowcharts to get any further. We had found at entry point, somehow, that there now had to be a solution to the main part concerning the mutual relationship between enzymes and RNA. For a long time, the starting point for my considerations had been the molecule that combined the two factors, the transport RNA. It just had to be the key molecule. It carried an exactly matching amino acid to the assembly location, and the information about it was located on the other side like a barcode on some food packaging. It is easy to imagine that today, one could place different loaded tRNAs next to one another, connect the amino acids on one side, and use each of the three bases (anticodon) on the other as a template for an attached RNA. This would result in the storage of a peptide in an RNA. This only works today because the amino acid on a synthetase is only linked to an associated tRNA, thus making the assignment of the anticodon unique. There were no specific assignments at the beginning. The way to solve the problem appeared to be to recognize a growing interdependency between the molecules, which could eventually lead to a corresponding assignment. A connection to the storage process when the peptides were formed must have existed from the very beginning. I was convinced that the coupling of the two molecule groups, the RNA and the peptides, had to be an iterative process, step by step: if you get one from me, I get one from you. And that with only a very few species, so that the possible combinations could not get out of hand.

The result of my deliberations leads to a hypothetical model that can be partially verified. Future experiments will show which is the most promising approach. In the following illustration, I attempt to take a new look at the problem of information storage on this basis. It represents the basis for the entire development of life. It proves helpful here to approach the crucial point of the first cell proliferation from two sides. One is the forward-looking one, which starts before life began, with the formation of molecules until the first successful cell division. The other is the backward-looking one that looks into the past from today's biochemical point of view.

The process that has ensured that living cells have grown and not died over the past billions of years must have had a beginning, a beginning that, with minimum equipment, provided the essential basis for information storage about the components that had been successfully designed until then. Many good reasons exist for moving this beginning into an open system in the earth's crust, in columns with a steady supply of molecules and the disposal of superfluous material. In anticipation of this, in my view, metabolism, cells, the equipment for a cell, and finally the multiplication of the overall system only can be developed under such general conditions.

But how did the formation of the first specific synthetases come about, which ultimately guarantee the reliable assignment of the building blocks needed in the ribosome for forming the enzymes and the synthetases themselves? Today, the code for their sequence is contained in the DNA, which can be read by the mRNA following transfer to the latter. As described above, a randomly formed RNA has no information value for the formation of amino acid chains. It only receives this value when it can be assigned to specific and coded amino acids that are required for building a functional peptide. The large number of amino acids in peptides, which can perform enzymatic functions, prevents any random combination that could occur in conjunction with the RNA. The possibilities for variation are so gigantic that the whole lifetime of the universe would not be long enough to get a useful result. At this point, it is my irrefutable conviction that a joint development must in principle have taken place that allowed the "software" (RNA) and the "hardware" (enzymes) to be mutually built up.

In order to make this understandable, I address the most important steps in the development of life, as they have emerged from the considerations made thus far, in the following chapters. They build on one another and sometimes took place in parallel to one another. This includes the sorting of the molecules (I), their selection process (II), the formation of an RNA subworld (III), the linking of the protein world with the RNA sub-world (IV), the beginning of information storage (V), and the process of the first cell division (VI). As a geologist, the first thing you learn as a fresher is to structure the earth's history and split it up into time slices. So why not at the start when everything began?

I beg for the reader's understanding when I frequently choose wording for linguistic reasons that sounds as if I were reporting on already known processes. It is and remains a hypothesis, so I should have written everything in the subjunctive. That is something I wanted to avoid, however.

8.2   Phase I: Formation and Enrichment

One of the essential prerequisites for the formation of larger molecules is the concentration and selection of raw materials from a wide variety of organic and inorganic molecules. A person who weighs 100 kg consists of some 10^{28} atoms. That is a one with 28 zeros. The number of molecules, composed of the atoms, is three to four orders of magnitude less. That means that on average we consist of maybe 10^{24} molecules. It can be safely assumed that in the earth's crust over more than 20 km in the vertical, a multiple of organic molecules can form—molecules that rise, come into contact with one another, and react to form new compounds. The number of dimensions alone makes it clear that an infinite variety of new chemical building blocks can be formed in these processes. Enrichments and sorting processes are constantly required to acquire something for forming biologically relevant molecules from the large cauldron of molecular soup. Corresponding operations exist in all natural environments where material transport takes place. The transport of sediment in a river from the mountains to the lowlands and finally to the sea can serve as an example here. Solely by being transported in flowing water does the rock material get separated by size and density. The biggest chunks remain in the mountains and have to be broken up into smaller chunks before being transported further down. The gravel and sand are deposited immediately when they arrive in the lowlands where the water flows slower. The clays largely remain in suspension and are transported to the sea. At certain points, heavy minerals such as magnetite, ore, or gold accumulate, even though there are only small grains of them. What does this have to do with a fault zone in the earth's crust? It has to do with fundamental separation processes that originate with flowing water.

In a deep open fracture zone, material flow is caused by gases and water rising toward the surface or in the opposite direction in the case water infiltration in higher mountain regions. As soon as communicating pathways appear in the form of fault paths that cross their way, the high-lying water entry points lead to an artesian outlet in lower-lying areas. Mass separations even occur in this fluid environment, but these are overlapped on the walls by mineral discharges on the surface. The molecules that move past the rock walls with the fluids can be held in place, build up multilayered layers, and bind other substances or allow them to pass through after reaction. The image of a bubble of air in an aquarium, which slowly rolls under an upward-sloping leaf, illustrates one of the many processes that take place at depth. The fracture zones often have sloping boundary surfaces under which small droplets of supercritical gas (carbon dioxide and nitrogen) roll upward. This is made possible by these droplets having a lower density than water. Molecular films, which adhere to the minerals in the walls to different degrees, come into direct contact with the interface to the supercritical gas bubbles. Substances can be absorbed here if they are more attracted to an organic solvent (hydrophobic molecules). If they are hydrophilic molecules that prefer to remain in the water, they may react with the hydrophobic molecules collected directly on the surface of the droplet and be carried away. Almost like building a snowman, the droplets collect a large amount of things that can be used later for more complex molecule formation.

The next step involves the separation or selection of molecules through recurring processes, in which the same processes consistently lead to similar groups of molecules over long periods of time (tens or hundreds of thousands of years and longer). The process described in the transition range from supercritical CO_2 to CO_2 gas (scCO_2 to gCO_2) at approx. 1000 m depth in the crust is a very effective process for creating vesicles (see Sect. 7.3). The blocks in the vesicles (incl. phospholipids) originate from the concatenation of carbon monoxide with hydrogen (similar to the Fischer-Tropsch synthesis) and, in the case of phosphate, from the mineral apatite. It is a mineral common in many rocks and is easily dissolved by acidic waters in the fault zones. In our high-pressure system under experimental conditions mirroring the top crust (1000 m depth), ground apatite was completely dissolved after a few days. Via the processes that take place in the 1000 m transition zone, amino acids that formed at different depths (incl. from NH_3, HCN, CO) can combine to form peptides and come into contact with the vesicles. The vesicles offer the chance to filter out certain amino acid chains, thanks to the composition and the structure of their membrane. Fluctuations in pressure caused by twice daily tides or even more frequently owing to geyser eruptions controlled by CO_2 lead to rhythmic changes that continuously control the formation and decay of vesicles and molecules. A particularly noteworthy aspect of the transition zone at a 1000 m depth is the generation of a large amount of entropy, both at the transition from supercritical CO_2 to gas (expansion, lower order) and vice versa (gas compression, warming, heat emission). It means that a reaction between the molecules can take place, although this creates order and reduces entropy, because of the large amount of entropy generated by expansion and dilution. At the same time, work is performed by pushing out the column of water when a geyser erupts. A small part of this drives turbulent circulations in the cavities. Certain chemical reactions in the micro-autoclaves also lead to the storage of energy. Reactions that occur in the environment of the scCO_2/

gCO$_2$ phase transition harbor ideal conditions for increasing entropy and converting energy. Changes in the pH value also take place, which varies due to fluctuations in pressure. In the supercritical phase owing to the high share of CO$_2$, this lies at about pH 3.3. As soon as the pressure drops and gas forms, the pH increases by about two units. At the same time, the expansion in the gas raises the temperature by up to 20 °C. After the eruption, returning water quickly causes the original pressure to build back up again. This compresses the existing gas up to the supercritical phase. The process is associated with a rise in temperature that at far above 60 °C reaches the melting temperature of a double-stranded RNA (see below). These special conditions at approx. 1000 m crust depth allow for reactions that were hardly possible on the surface of infant earth. Drill cores acquired from deep holes at a depth of 1000 m in mofettes (CO$_2$ geysers) may still demonstrate these processes today.

The range of molecules in this process was and still is gigantic. Twenty amino acids are used predominantly in the biological cells. However, over 400 different species are known in the meantime, most of which play no role in nature. Studies on the temporal appearance of the amino acid species in the cells have shown that an older group of about 10 to 12 different ones exists that can be formed hydrothermally [1, 2]. Except for glycine, the simplest amino acid, they all have at least two different structures (D and L), in which the chemical composition is identical. They are chiral and have either a left- or right-handed configuration. Which selection processes ultimately led to only the canonical amino acids (those used in nature for the construction of proteins) and those with exclusively the L-handedness, with only a few exceptions, being used in evolution?

But stop! A certain amount of D-amino acids also exists in the cells, which can vary depending on the type of cell. Two main reasons exist for this. The first is physicochemical. The goal of a group of molecules is always to assume an energetically favorable state. This is then achieved for an amino acid species when the ratio between the D and L versions is 1:1. That means that an initial quantity of 100% L-amino acid transforms into an amount of 50% D- and 50% L-amino acid in a characteristic time period (which can vary greatly depending on the species)—an effect used to determine the age of organic substances such as is used in forensics.

The second cause lies in an alternative possibility of forming certain peptides, in which atypical amino acids such as D-amino acids are incorporated. They are not linked to a ribosome that controls the exact assignment of the amino acid-loaded tRNA through the usual process and only allow the L variant. The formation takes place in large enzyme complexes without the involvement of an RNA (non-ribosomal peptide) [3]. This complex can be thought of as a three-dimensional puzzle made up of various enzymes, each of which was formed individually on the ribosome in the usual way. In this construct, a completely different type of peptide is catalyzed completely independently of a ribosome and a template from an RNA, whose structure is not stored in the DNA. Its blueprint is derived indirectly over bands, only through the specified arrangement of the enzymes in this complex, which controls the assembly through precisely coordinated, successive steps. This type of peptide formation occurs in some types of bacteria, archaea, and fungi but also in various snails living in the sea.

What use does this information have in relation to our consideration of molecular sorting? Well, it shows that no biological law exists prohibiting the use of D-amino acids in the cells. Their existence in some peptides may provide an indication of the reasons for the exclusive determination of the L-amino acids in the proteins and enzymes in the course of chemical evolution (chirality problem). The D species are incorporated in the puzzle-like enzyme complexes without the involvement of an RNA, only through defined reaction steps that are catalyzed exclusively by the enzymes involved. This means that the enzymes, although they are all made up of L-amino acids, do not impose a mandatory definition for one or the other orientation of the amino acids in the peptide to be newly formed. The situation is completely different when peptides are formed on a ribosome. Linking a D species has absolutely no chance in this relation today. In addition to amino acids, sugar and organic bases can also be derived as products from the hydrothermal environment. Just like the amino acids, they are represented with numerous species, of which only those that are present in RNA and DNA have been selected. Ribose, the sugar in RNA, is only used in its right-handed form (D), just like its relative, deoxyribose, in DNA. In addition, various L-type sugars also exist that are used in the processes in biological cells.

In summary, the following can be stated in relation to a fluid-bearing fault zone in the crust:

- Organic molecules can be formed vertically over long distances in the continental crust. Chiral molecules always produce the same number of L and D species.
- Rock minerals' dissolving releases phosphate, metals, boron, and other substances.
- A process of collection caused by rising supercritical gases leads to a strong accumulation in cavities (micro-autoclaves) and in the transition zone to the subcritical gas at approx. 1000 m.

- Optimal conditions exist with regard to energy and entropy which favors a linking of the molecules into chains. This has already been confirmed in laboratory tests.

8.3 Phase II: The Selection Process

When imagining how extensive the range of molecules was, the question arises immediately as to which physicochemical laws must have ultimately guided the selection processes when life began.

The vesicle model (Sect. 7.3), about which we have already gained laboratory experience in terms of molecular separation, can serve as a starting point here. The vesicles that emerge in the $scCO_2/gCO_2$ transition zone are not all the same. Depending on the starting materials, a wide variety of lipids are formed, each of which forms different structures in the membranes. Different vesicle shells interact differently with surrounding peptides, which means that a first molecular separation takes place solely on the basis of the membrane structures. But all vesicles work toward a separation into hydrophobic and hydrophilic amino acids. The separation takes place in the cavities through the coexistence of water and supercritical CO_2, which serves as a solvent for hydrophobic molecules. A chain of hydrophobic and hydrophilic species at the interface results in peptides with amino acids, which have both properties. Depending on which section of the chain predominates, the peptides may turn out to be hydrophobic or hydrophilic. If the organic solvent disappears because it suddenly becomes gaseous, the hydrophobic sections of the amino acid chain can "hide" in the vesicle cell walls, while the other sections look outward. If turbulence subsequently destroys the vesicles when there is a loss in pressure, the peptides are forced into the water.

Then the following happens: The long chains try again to keep the hydrophobic sections in their union away from the water. They do this by structuring themselves, folding themselves into complex structures, and as far as possible surrounding the hydrophobic sections with the hydrophilic ones. The result is a kind of ball with a hydrophobic center and an envelope that is hydrophilic. At this point, the same supply of molecules with the L and D configurations can lead to the chains of amino acids also having a complex mixture of both forms. In evolution, only the L form prevailed for amino acids at a certain point in time. It is therefore important to identify a process that led to a separation of both orientations. It was only when peptides that consisted of only one species existed (enantiomerically pure peptides) that the race between the two versions was able to begin. It is not sensible for this reason to rely on random links that may have led to an enantiomerically pure peptide. Infinitely small probabilities for their formation come into play very quickly with this approach, which due to the high number of amino acids and possible variations (see above).

No, a separation must have taken place that was dictated by the environment. And so, we decide to take another look at the situation in the fault zones. Studies on peptide formation have shown that with an amino acid mixture with both handedness in equal parts (racemic) in a neutral environment at pH 7, chains are formed that incorporate both variants. These are not enantiopure. However, if acidic conditions exist with low pH values as in the hydrothermal crevices, peptides are formed with predominantly only one handedness [4]. This important finding cannot be overemphasized since it is an important prerequisite for molecular separation. It represents one of the essential foundations for further development, which immediately becomes clear when we take a look at the continuing steps. From a certain length, the chains quickly form three-dimensional structures. They are significantly more stable than unfolded peptides and therefore have a longer lifespan.

However, what order of magnitude do we mean when it comes to the lifespan of peptides? A simple question to which the answer is complex: again, it depends on the circumstances. Subjected to high pressure and high temperature in the earth's crust, we must assume a lifespan of hours, maybe days, at low pH values. At temperatures below 40 °C and in neutral conditions, the lifespan increases to several hundred to over a thousand years, depending on the type of amino acids.

Thus, we have identified a process specific to the early days of the earth that led to an effective separation of right- and left-oriented amino acids as soon as peptides were formed in an acidic environment. In this phase, however, it had not yet been decided whether right- or left-handedness would prevail later. The choice of the canonical amino acid species used today also took place later.

The description of an amino acid selection so far has focused on differences which are given by the hydrophobic and hydrophilic properties of the species in question. Because of the low pH values in the hydrothermal fault zone, the peptide ball with its hydrophobic center may have consisted of either only amino acids of the L or D form. Regardless of the configuration, the amino acids can still be distinguished by whether they prefer to adopt or give up H^+ ions, i.e., whether they react acidic, basic, or neutral. These properties also need to be considered for further selection processes. This is because, depending on which amino acid type dominates in a peptide, differences in the properties of the entire peptide exist. Among other things, this comes into play when the pH values and thus the supply of H^+ ions change during the phase transitions at 1000 m depth. The relief in pressure when a geyser erupts causes fluctuations in the pH value of up to two units. Overall, the selection process coupled to the vesicles leads to longer enantiomerically pure peptides over long periods of time, which can fold into complex structures. Depending on the type of structure, the possibility exists that the peptides become catalytically active. This creates a relatively simple molecular tool with enzymatic functions.

We have to accept at this juncture that the development up to this point in time could have produced complex molecules but that the enzyme-like structures were nothing more than one-hit wonders. When they hydrolyzed, i.e., were broken down into their constituents, no identical molecule moved into their place. Each peptide formed in this way was a random product that could not be reproduced in the same form after it had decayed. The building plan was disposed of, and the mechanism that made it possible to precisely reproduce it went missing. Nevertheless, we can still gain something from this process. A basic pattern can be recognized from the vesicle/peptide interaction, which always leads to similarly structured peptides owing to the respective structure of the vesicle membrane, a type of selection that in the further course is important.

It turns out that structured peptides with amino acids with the same handedness could be formed by the rhythmic formation of vesicles in an acidic environment and in large numbers. It must be taken into account here that there was probably a strong imbalance in the supply of the individual amino acids. The most frequently represented species must necessarily have been those with the simplest structure and the most favorable conditions for formation, therefore. These are glycine and alanine. Taking into account the large numbers of possible combinations, it is again fortunate that glycine is achiral. Under acidic conditions, enantiomerically pure peptides can easily be obtained in combination with alanine, which have either an L or a D configuration. A condition for this is clearly an oversupply of these two simplest amino acids. The other eight to ten hydrothermal amino acids appeared with decreasing concentration corresponding to their complexity. The other half of the 20 canonical amino acids probably only took part in the further development of the cells at a later stage, probably in conjunction with already more highly developed enzymes.

From today's perspective, groups (still as L and D forms) can be separated from the large number of folded peptides which had different structures or concentrations of certain amino acid species (Fig. 8.1). In addition, in total, they also behaved more neutral, basic, or acidic. Firstly, the grouping was determined in the course of development by the structure and composition of

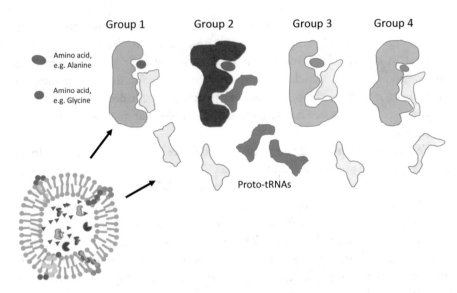

Fig. 8.1 Formation of groups of similar enzymes, partially in interaction with lipids in the vesicle membrane (Groups 1 and 2). The sequence of the amino acids is not stored, and it contains only relatively unspecific information, partially based on the different lipid membranes of the vesicles. Other enzymes also exist that are freely formed, without interaction with vesicles (Groups 3 and 4). The enzymes form weak specific bonds both with proto-tRNAs and with certain amino acids. They catalyze the linking of the existing amino acids with the tRNA (preferred pairings with the most frequently represented amino acids glycine and alanine). The weak specific contacts do not result from the sequence of the amino acids in the enzymes but from the affinity for the group type

the vesicle membrane and secondly by formation processes that did not interact with the vesicles. Owing to the fact that similar vesicles were always formed, and the other general conditions remained similar, similarly formed members of the groups always existed. Even today, certain functional molecules can be differentiated into at least two morphologically different variants. These include the tRNA synthetases, in which, for instance, differences in the size of the recognition structures for the attachment of the tRNA [5, 6] exist. This point will continue to be an important argument in the discussion about the storage of information in RNA.

The tRNA does not carry any specific information at this point in time. The constant copying leads to many variations caused by mutations. This mainly affects single-stranded sections such as those of the anticodon or other loops in the molecule. The different variants have different priorities for contacting the enzymes. Not every tRNA connects with every group type. Ultimately, it is two systems coupling that have only very weak specific information. A specific assignment and the storage of the information in an RNA would need to develop from this with the next development steps.

In summary, it can be stated that:

- Vesicles and peptides form in the crust at a depth of approx. 1000 m.
- Certain peptides interact with the vesicles; the peptides that are protected by the membrane accumulate.
- Different vesicle membranes lead to differently structured peptides, which are highly variable in the order of the amino acids but can be assigned to groups. Peptide groups also exist that do not interact with vesicles.
- Mixed peptides with L- and D-amino acids are normally formed. However, differences exist that depend on the pH. While the mixed peptides with both orientations are formed in the neutral range at pH 7, enantiomerically pure one can only occur from one species at low pH values.
- If longer chains form in this environment, three-dimensional structures develop quickly that are more stable and survive longer than unfolded peptides.
- The hydrothermal environment leads to the preferred formation of the simplest amino acids glycine and alanine. Glycine is achiral and alanine is chiral. Under acidic conditions, this combination can easily form enantiomerically pure peptides from only two amino acids.
- A tRNA develops from the ribosomal RNA, which through mutations forms numerous variations. The different formations have different preferences in contact with the enzymes from the different groups.

8.4 Phase V (Brought Forward): A Possible Start to Life

The explanations in the two previous chapters are supported by experimental evidence concerning the reactions of peptides on vesicle membranes under the conditions in the deeper crust. They form part of the research results from the Essen research group (see Sect. 7.3). Older work exists for classifying groups of today's synthetases and associated amino acids, e.g., Eriani et al. [5]. In the following chapters, I use the knowledge and results so far available as a basis for designing a purely hypothetical scenario for continued development. It aims to show one of perhaps many possible ways under realistic conditions for how complex reactions and processes may have led to the formation of LUCA. The scenario is also supported by data and proven coherences that represent the latest scientific findings.

The situation described in Phase I (formation and enrichment) and Phase II (selection process) shows that vesicle formation is of particular importance with the chemical evolution of peptides. As a consequence, with the formation of cavities in the continental crust, geotectonic conditions contributed to durable selection processes that led to an abundance of presorted molecules. In order to be able to better understand the continued development steps, it is helpful to approach from the other side first, from the side where life already exists. If we go backward in time from the complex situation we have today toward the origin, the question arises as to from which point in development we can actually speak about the beginning of life. However, the fundamental problem exists that, as Sect. 1.2 described, there is no definition of life as such that is universally valid and accepted by all scientists. This is partly due to the perspective taken by the different disciplines. Whereas physicochemists aim to combine order, energy, and entropy, the biologists are more concerned with information transfer during replication, metabolism, the formation of compartments, regulation, and the other essential properties of life. Further definitions refer to aspects of self-creation and self-preservation in a closed system. They all show that the attempt to define life is sketchy and controversial. From this starting point, it becomes clear that a definition of the beginning, which until now has not even been made concrete, seems almost impossible. Nevertheless, within the framework of the hypothetical model presented, I will attempt to narrow down the initial development of life and discuss a possible start. The key characteristics defined by biologists from the perspective of existing life should serve as a basis for this and can be supplemented by a new approach, as described in Chap. 9.

Now that we are approaching the decisive point in the origin of life going backward, we need to skip Phases III and IV, which follow Phase II in chronologically terms. These time slices hide beginnings that are easier to recognize if the end result has already become visible. The corresponding Phases III and IV, which show the key process for data storage, are considered in the subsequent chapters that follow Phase VI.

In Phase V, we immerse ourselves in a time that, from a model perspective, is already after the decisive breakthrough in the reaction processes for the molecules. I define the breakthrough as the occurrence of a molecular group in higher concentrations, which consists of two transport RNAs (tRNAs) and two enzymes (synthetases), which have the ability to specifically load one of the two tRNAs. This also includes an RNA that is enzymatically active and corresponds to a precursor ribosome (rRNA). The information from the two synthetases and the two tRNAs and the rRNA is stored in an RNA. This RNA is the information store, whose function is initially taken over by the DNA much later. The two synthetases are—and this is special—each composed of just two amino acid species (glycine and alanine), but each has a different sequence and size. The tRNAs are loaded with an amino acid from one synthetase with glycine specifically and from the other with alanine. At this point, everything exists in large quantities in the open system of a fault zone in a wide variety of cavities. It may not be immediately apparent, but such equipment allows the principles of life to begin to take hold for the first time.

Before taking a closer look at this particular state of development, it is worth trying to compare the technology again, ideally from the playful perspective of the world of Lego. We are gazing down on a small factory in which two different sized robotic machines can be found in operation. They are made up of a combination of just two standard bricks from a Lego set that only has six and eight sectioned bricks. Even the color of the components (red and white) is the same for both, so that the robots only differ in the order of the components used and in the size and the task of the work to be performed. The small robot's task is to repeatedly pick out the red six-sectioned bricks from a pile of colorful Lego bricks and place them on a suitable transporter. The large robot has the same task but for the white eight-sectioned bricks. The myriad loaded transporters each have a barcode as an ID which is specially customized to the goods being transported. They approach a machine which recognizes them from their barcode. The machine continuously assembles new robots from six- and eight-sectioned bricks based on a blueprint from an engineering office. These are the two robot types that loaded the transporters at the beginning. To do so, the right components need to be selected in the right order to guarantee a precise assembly. Depending on the requirements, the machine allows the associated transporter to approach an assembly line from which the six- or eight-sectioned bricks are then unloaded. There, the new component is attached to the parts which have already been assembled in advance, in exact accordance with the blueprint. In this way new robots are built continuously as long as the necessary components are supplied. Following completion, the newly built robots take on the same tasks as the older robots. To avoid bottlenecks in transportation, new transporters have to be continuously built in another part of the factory and assigned the appropriate barcodes. This is the image of a self-building system that, if sufficient quantities of components are supplied, maintains itself and also increases in size. The robots build themselves.

In a fault zone, the process is understandably much more complex but could look something like this. In the abundance of molecules, the autoclaves offer, the synthetases (our two robots) repeatedly succeed in loading their specific amino acid (glycine or alanine) onto the corresponding tRNA. The transporters arrive at one of the RNAs through a weak flow and sit on it in the order speci-fied by the codons. Assistance is already provided by an enzymatically active RNA (rRNA), which can be regarded as a kind of precursor ribosome (P-ribosome). The amino acids are linked in sequence with the assistance of the P-ribosome and thus reproduce the synthetases which the RNA memory pre-determined and which preceded the cycle. The RNA itself, which also contains the information from the tRNAs, can also be copied by adding complementary nucleotides. To do so, an intermediate step is required, which takes place through the formation of a complementary strand (negative impression), from which the renewed complementary attachment then adopts the sequence of the bases. This ensures that all the essential building blocks of the cycle—synthe-tases, tRNAs, P-ribosome, and the information-carrying RNAs—are preserved. And all components multiply as well because the replication takes place faster than the decay.

The description of the processes in the cavities presents a model that has far-reaching consequences for the discussion about the origin of life. It is therefore important to look at the individual steps in the chain of argumenta-tion in detail in order to understand which factors have come into play. As before, we are still talking about an open fault system in the continental crust in the transition phase from supercritical to subcritical CO_2. And it continues to concern the idea that the development that led to the first life-like process did not take place in a cell but in a network of larger and smaller reaction chambers. The cavities, which act like autoclaves, could have the diameter of biological cells today (micro-autoclaves) but could also form fist-sized or per-haps more flat, large volume bodies. The open system offered the advantage of constant molecular replenishment and the greater abundance of reaction part-ners, without the need for a developed gatekeeper system for a separate cell compartment. The considerations lead to a principle approach that can only be presented in a simplified way. Individual physicochemical steps for this cannot be yet discussed right down to the last detail, since the range in varia-tion and the number of reaction partners involved go beyond our imagina-tion. What it concerns, however, is a principle that could have prevailed from a certain point in time and continues to this day in the living world.

It is easy to understand that the functional molecules involved in the early phase, i.e., the vital components of a cell, were not as complex then as we find they are today. LUCA, the cell from which all cells originate today, was one of the seedless prokaryotes. It was only billions of years later that eukaryotic cells with a nucleus developed from them. As with the first car, the first self-prop-agating cell must have had a minimum of equipment from which everything else developed. Before there was an accumulation of all the vital units in one envelope, a minimum number of components must have existed that could start a functioning information storage system (based on the extracellular development model presented).

The open system allowed for the continuous supply and removal of building blocks but also temporally stationary conditions in protected autoclaves, from which a limited exchange with other spaces was possible. I will discuss how the functional molecules were formed later. At this point in time (Phase V), the RNA had developed to such an extent that it already carried the information from the synthetases, the transport RNA (tRNA), and the RNA from a possible P-ribosome (rRNA).

According to the model, the two synthetases (S1 and S2) are each made up only of the same two amino acid species A1 (glycine) and A2 (alanine) (Fig. 8.2). However, A1 and A2 each vary in number and order, so that the synthetases (our robots) have different catalytic properties. For example, S1 can specifically link one of its own amino acids (A1, glycine) to one of the tRNAs, while the second synthetase (S2) specifically loads the other amino acid (A2, alanine) onto a second tRNA. This works because the synthetases have already developed a recognition structure for binding a suitable tRNA (R1 or R2). (Alternatively, there could have been several tRNAs, which reduces the accuracy of the assignment. The example here concerns the minimum equipment.)

After loading the tRNAs (of which there are many copies that are constantly loaded on the synthetases), these come into contact with a precursor RNA (proto-RNA, the blueprint from the engineering office), which contains the entire code for the synthetases (S1 and/or S2). The proto-RNA codons provide the tRNA anticodons with the appropriate docking sites, which the

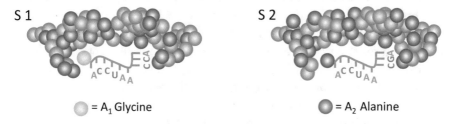

Fig. 8.2 Two different synthetases, consisting of the same amino acids glycine and alanine exclusively. S1, in specific contact with a tRNA, to which glycine is attached, and S2, with a tRNA, which is loaded with alanine. The achiral property of glycine favors the formation of a chain with a second, chiral amino acid, in which only D or L configurations occur (enantiomerically pure chains, only a schematic representation of the tRNA). The information about the order of the amino acid species is already stored in an RNA at this time

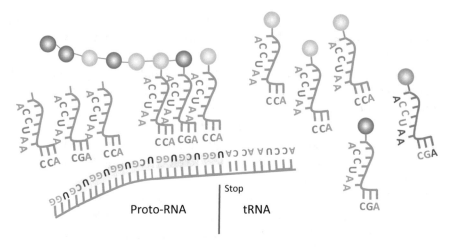

Fig. 8.3 Proto-RNA code for two amino acid species with the code for one tRNA attached. At the same time, the tRNA performs the stop function in the selection process during the translation, since conversion to a peptide does not take place owing to the lack of a specific amino acid. Length of the tRNA symbolically. The first RNA, which has saved specific blueprints for two enzymes (proteinogenic RNA), will be used in the further development as a template for the continuous production of the first two specific synthetases. This is possible because the information can be selected from the RNA (which is now comparable to the later mRNA) through the tRNA's anticodons

respective tRNA can now briefly occupy. The P-ribosome provides assistance for this. Contact is maintained until the neighboring triplet is also adopted by a loaded tRNA with a suitable anticodon. At this moment, the amino acid for the first tRNA is linked to the one that follows (Fig. 8.3). The discharged tRNA then leaves the strand and can be loaded back onto the associated synthetase. In the meantime, many more tRNAs have been loaded and have reached the RNA strand. They join the process and link the amino acids they have brought along with them as specified by the information store until the end of the RNA for synthetase S1 is reached. This is a process that takes place in cells in a more complex form today. The causes for all reactions lie in the physicochemical laws according to which energetically lower levels are adopted under increasing entropy (e.g., through heat dissipation).

At this moment, the strand of information store RNA (later mRNA) is fully translated into the sequence of a chain from the two amino acids used. The same takes place on a second RNA (for S2) or in parallel with the linking process for the first S1 part section. In this case, the RNA must be present as a coherent strand for both synthetases (S1 and S2). However, for this to take place, an intermediate building block is required after the S1 section, which acts as a stop element and stops the amino acids linking. The same applies to the RNA section for S2. The entire process continuously produces synthetases with exactly the same composition as that of the first two synthetases. Thus, they do not die out but multiply. Prerequisite for this is the constant replenishment of the molecules and a replication of the RNA, which carries the information for the chains. If the RNA did not replicate, its information would be lost following the decay.

A process begun in this way, which leads to the preservation of the most important building blocks of life, can be seen as a forerunner for the principle of life—but only in an open system. The driving forces for this process are formed by the cyclical relaxation of the supercritical CO_2 while increasing the entropy, the flow turbulence, and the heat transfer processes. We have identified the geysers controlled by CO_2 gas as the motor that ultimately started the processes for the development of life, which in turn are driven by the energy stored inside the earth. The sum of the physical processes in a geyser system of this type is surprisingly comparable to the processes to be found in a simple steam engine. But more on that later.

So far, the process is almost identical to the processes for forming synthetases that take place in cells today. The main difference is that certain components, such as tRNAs, amino acids, and synthetases, are limited to two units each. The replication of RNA and the linking of amino acids are controlled with the help of enzymes today. They facilitate a reaction in the aqueous environment of the cell and accelerate the process considerably. Physicochemical laws dictate that the reactions can also take place without enzyme catalysts. However, the amount of time they need is correspondingly longer, and there is a risk that the decay takes place faster than the molecules can form. In addition to the support provided by a catalytically active P-ribosome, some of the reactions shown here may also possibly occur directly in the supercritical CO_2.

If we go back to the example with the Legoland robots, we still haven't progressed one step further. It is the same robots that are constantly being produced. There is no new development. At this point, we can sit back and observe how the system with the two synthetase variants, the two tRNAs, a simple ribosome, and an RNA is kept alive for whatever length of time. The prerequisite for this is that the geyser process functions permanently and that there is an adequate supply of building blocks for forming the molecules. If this is the case, the community of molecules concerned grows until the availability of raw materials sets a limit. As exciting as it is, however, this process does not go much further. We need a process to come into play that controls the continued development of molecule formation and increases the complexity of the entire system.

If we imagine the open system of the cavities with all its mass transfer and reaction processes, we can guess that more was possible than just copying synthetases S1 and S2 using the existing RNA. At the same time, the possibility must have existed for specifically loaded tRNAs to be attached to one another, either with the help of the P-ribosome or without an RNA strand template. In these cases, the amino acids glycine and alanine brought along were linked to randomly sorted chains. This did not achieve much, however. But it did open up a new opportunity that gave strong impetus to the continued development.

Let's have a precursory look at the world of Legoland again in relation. In the factory, more and more transporters are being loaded by more and more robots, but not all of them reach the assembly line. A hall exists where the transporters are lined up in random order and have to hand over their Lego bricks. Once they are handed over, they are put together to form wild, random constructs and released into the open once they reach a certain size. At some point, they disintegrate again, and the individual components are placed on the pile. This is actually a redundant process that just uses up resources. But there are enough of them. From time to time, however, some of the line of transporters scan the barcodes and attach them to the blueprint for the robots in the office. That has consequences for the production. In the future, not only will the robots be assembled from the now supplemented blueprint but also the randomly assembled constructs with no function. The process takes a very long time with absolutely useless random constructions in all variations being built time and time again. Many of them find their way into the blueprint through the scan process. But suddenly a version is included that creates a new robot. This robot is also assembled from the red six-sectioned and the white eight-sectioned bricks. But this one is different from the other two. It can place another Lego brick, the green four-sectioned one, onto a new transporter. But the thing is that this transporter is not even needed. This means that transporters carrying the four-sectioned bricks have no place in the machine that builds the robots. According to the blueprint, only robots made of six- and eight-sectioned bricks are built here—but now there are three different types. But this gives rise to a new possibility. In the hall where the surplus transporters are lined up, more and more transporters are now loaded with green four-sectioned bricks on top of those with white and red bricks. These come from the third robot. This results in a completely new mix. As before, the bricks are combined to form wild constructs without any function. Some of them are scanned again and attached to the blueprint from the office. And then, after myriad constructs, one is formed that results in a new and functional robot. It is built out of white, red, and green bricks. The

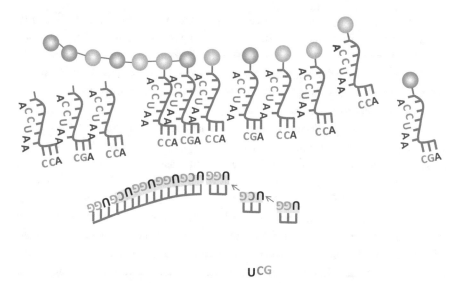

Fig. 8.4 Example of the formation of a peptide with a random sequence of two amino acid species in free space. Simultaneous documentation of the sequence by attaching nucleotides and formation of a new RNA from the tRNA templates (shown here without precursor ribosome)

barcode for this is scanned and attached to the main blueprint, meaning it can be reproduced like the other robots. Its ability lies in being able to load a blue, two-sectioned brick onto a passing transporter....

Now let's go back to the cavities in the fault zones: with each randomly arranged linking step, the tRNA's anticodons lie side by side for a short while and form a template that can be supplemented by complementary nucleotides (Fig. 8.4). If the complementary nucleotides combine to form an RNA chain, then the information about the sequence of the randomly linked amino acids (which almost exclusively belong to the wild constructs) is stored in it. In other words, every random sequence of amino acids in the resulting peptide has the chance of being simultaneously stored in an RNA that occurs in parallel.

Is this not what we are looking for in relation to the construction of proteins to understand information storage? It is certain that storage in an RNA only occurred in a fraction of the possible cases. However, we did have an enormously long period of time during which a small number of successful documentations in each cycle would be sufficient to produce a large number of new molecules where the amino acid sequence was stored in an RNA. The amino acid chains randomly linked in this way had no function. But they did harbor a great potential, since they appeared with an almost unlimited variety of variations. Saving their sequence in an RNA formed in parallel offered the chance for them to be formed time and time again without the chance of getting lost. If some of the proteins were later discovered to be functional carriers, then they had the chance to assert themselves immediately. In my opinion, this situation best explains the requisite rapid development of the proteins as function carriers for future cells.

In the event that a P-ribosome assisted in the random link, it means that the attachment of the nucleotides to the anticodons of the randomly arranged tRNAs was not left completely free. The link was made accordingly more or less specifically by the catalytic function of the simple P-ribosome. This may mean that there might have been at least two different P-ribosomes. One catalyzed the formation of the synthetases by selecting the information from one of the early RNAs, and the other catalyzed the formation of random amino acid chains without any information. The random formation of the amino acid chains means that the documentation of the random sequence in an RNA would have been possible at the same time. Or maybe only one ribosome molecule existed that was capable of performing both functions.

In the further course, the free combination of tRNAs may have been restricted or even prevented, perhaps after the molecular development had just taken place in only one cell, meaning that the formation of new RNA on the ribosome was then no longer possible. Instead, the existing RNA strands could be used as information carriers for linking the peptides, as is now performed exclusively in the ribosome (peptide synthesis, in which the mRNA in the ribosome serves as an information carrier for linking the peptide chain). Today, the incoming tRNAs have to wait until they are approved for the passing code on the mRNA. Only then is the amino acid that is brought along linked. In the other case of random ordering postulated, the incoming nucleotides had to wait until they could occupy a place on the tRNAs' anticodon.

Each time a peptide with a random amino acid sequence is formed, a representative for an enormously large number of variations is created, almost 100% of which have no functions. They get lost again if they are not stored with their sequence in an RNA in rare cases as described. Proteins that fold and develop into catalytically active enzymes can exist among these. A functional enzyme would, for example, be a third specific synthetase which allows a further tRNA to be loaded specific (Fig. 8.5). The third tRNA may have resulted from mutations in which a base was exchanged in the anticodon. Together with the new synthetase, this would result in a third amino acid (the green Lego brick), which brings about a further increase in the possible

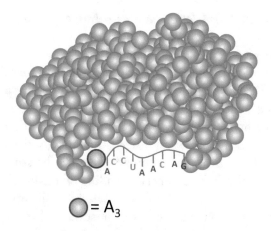

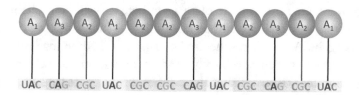

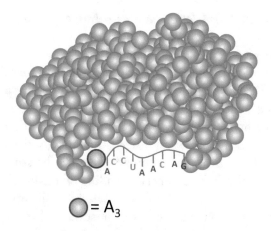

Fig. 8.5 A newly originated tRNA synthetase from a random combination of A1 and A2, which leads to a specific binding for A3 and linking to a third tRNA. A third amino acid comes into play with this third synthetase, which can be loaded specifically onto a tRNA. This increases the possibility of variation with further random links, so that a fourth and fifth synthetase, etc. can be formed in the same way over time, whose information about the sequence can be stored in a parallel RNA

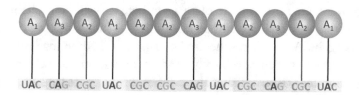

Fig. 8.6 Example of a chain with three different amino acids that were specifically loaded on tRNAs (tRNAs only represented schematically with anticodons). A total number of 531,441 different combinations already results with 3 amino acids and 12 units

variations when the tRNAs are randomly combined (Fig. 8.6). The interesting thing about this process is that the new synthetase still consists of only two amino acid species (glycine and alanine) but that a third species links precisely to a tRNA. If the information about the sequence of the amino acids for the new synthetase is stored in an RNA formed in parallel, this can, like the third tRNA, be attached to already existing RNAs. The grown RNA is now available for copies of all building blocks, including the third synthetase.

This means that we can determine the following for the development up to this step: information from three synthetases (still made up of two amino acid species) is available in an RNA that can specifically link three amino acids to three different tRNAs. Furthermore, information from the three tRNAs is contained in the extended RNA (and as before the information about the P-ribosome).

The piles of paper we had scribbled on had paid off at the latest at this point in our search for a solution. We slowly have an inkling that the chicken and egg problem had to be solvable, on paper at least. At that moment, the potential combinations which result from the method described suddenly became clear. On the one hand, we had the random formation of a large number of proteins, with their structure being stored simultaneously in an RNA, and on the other, we had a chance of obtaining further synthetases from these proteins, which could also lead to a specific loading of a tRNA.

An enormous increase in the possible variations in the free combination of the tRNAs takes place once again with the successful formation of a third synthetase, which is available for the third specific loading of a further tRNA. With the potency of 3^n, linking three different amino acids in a chain of 15 units results in more than 14 million different possibilities. The composition of the third synthetase is stored in the RNA at this stage of development, so that it can be reproduced continuously. The free assignment of loaded tRNAs with three amino acid species to form peptides is possible from now on. The same process takes place again here as well. New chains are created almost all of which are completely unusable, but also some can take on functions again. A fourth synthetase exists here, this time made up of three different species, which specifically loads a fourth amino acid species for another (mutated) tRNA. Again, this synthetase can be stored simultaneously in an RNA. Now anyone can work out easily how things proceed. The next variations are becoming more and more complex, with a maximum of as many amino acids as are available in the environment.

At this point, it may be appropriate to make one thing clear: in this phase, two developments occur in parallel side by side. One development guarantees the preservation and multiplication of the synthetases and the tRNAs, RNAs, and the P-ribosomes, while the other uses the possibility of infinite combinations for the formation of new peptides, whose information can be stored simultaneously in an RNA. If the storage fails, the peptide has no future. With this construct of a life laboratory, you get an idea for the first time of how the highly complex molecular compositions could be tested in infinite variations and how they could be stored through the RNA. The large number of possibilities explains why not all variations were used. It sufficed if a variant that was simultaneously stored in the RNA could take over a function in the complex system. Then the principle of selection took hold, and the molecule was given a future.

8.4.1 How Can We Imagine the Situation in Spatial Terms?

Depending on the height of the water column ejected, the pressure fluctuations that result from a gas-driven geyser eruption affect a crust section in several hundred meters depth. This means that below the boundary of around 800 m, perhaps as far down as 1100 m, all existing supercritical autoclaves become subcritical each time. This causes turbulent mixing. Cooler water flows from the sides from higher positions in the direction of the channel of ascent, and supercritical gas droplets rise from the deep to the boundary zone. Lateral currents and paths crossing lead to the spread of the newly formed molecules. Once started and after the information stored in the RNA is presented, the molecular concentrations are enough to continuously produce synthetases, copy the RNAs, and try new random amino acid sequences within the peptides being formed. The system is dynamic and extremely durable. If, in our minds, we give this molecular soup several million years, it is easier to understand that a great many variations could be attempted, and this is on tens of thousands of cold water geysers at the same time, which, owing to the initially high gas quantities from the mantle, were certainly distributed in large numbers along the fault zones. In the end, it was enough for the combinations of molecular formations and links to be successful at one point for a breakthrough to be achieved.

But there is still a problem to be solved in all of this: The production of lots of enzymes without a function led to the possible storage of data containing useless material. However, this data was reselected in the same way as that in enzymes with function which were used later. How were the useful synthetases selected from the pool of all enzymes?

One explanation could be that only a fraction of all randomly formed peptides were stored simultaneously in an RNA, which is the most likely case. Firstly, a large number of noncoding DNA segments exist in the genome of the eukaryotes, which means that they are not needed for forming the proteins. After all, it is 95% in human DNA. Some of these noncoding sections contain the template for the tRNAs and the ribosomal RNA. Nevertheless, a large percentage of DNA cannot be assigned a function. This is referred to as junk or scrap DNA. This percentage has decreased in recent years as more and more parts of the so-called junk DNA have been recognized as being necessary for certain processes. However, a large part still seems to have no meaning. Two opposing camps on this topic exist among scientists now. And the situation is far from clear. Perhaps one of the causes is the storage of the information from the peptides which have arisen at random. The proportion of noncoding DNA in bacteria and archaea is significantly lower. It lies at under 20%. The representatives of these domains have a rather circular DNA, in which the presence of useless DNA would quickly become disadvantageous. The lines of development for bacteria and archaea separated early from those for the eukaryotes. Perhaps shortly afterward they developed a way ridding themselves of most of the redundant RNA/DNA. This process would not have taken place in eukaryotes line, the DNA of which is rope-shaped with a complex winding structure. Another possibility is that the junk DNA has found its way into the eukaryotes secondarily because systems for defending against retroviruses had failed. The retroviruses embedded themselves in the DNA, multiplied, and finally took on functions. It is certainly exciting to look at the existing questions with a new approach.

The game of randomly combining loaded tRNAs for peptide formation can now be repeated up to the 20th synthetase. However, it can be assumed that with the increased entry of enzymes into the overall development, other processes contributed to the formation of the synthetases, so that the path shown would apply predominantly to the initial phase with amino acids from the hydrothermal environment. In any case, this explains the formation of the most important enzymes with simultaneous information storage in an RNA in a relatively short period of time. In addition, a huge selection of other peptides and enzymes can be explained, the structure of which was sometimes saved in the RNA even though they had no function. The nonfunctional sections could possibly be modified by mutations in a later phase, so that enzymes were formed which were available for new requirements.

8.5   Phase VI: LUCA Becomes Visible

In reality, the process of synthetase formation and the appropriate assignment to a tRNA could have continued up to at least 60 (variation of three codons possible from the four bases: $4^3 = 64$ options, minus 4 function codons). This did not take place. As an alternative, a multiple assignment of the tRNAs has taken place, which makes up to 6 "code words" for an amino acid (e.g., serine is encoded over six different base triplets). This stop can perhaps be explained by a fundamental change in the development of the entire process. It could mean that from a certain point in time, with a minimum number of 20 synthetases today, a transition took place from the wide space available in the autoclaves to the narrow confines of the cell compartments. As described in the experimental procedure for vesicle formation, many molecular building blocks from the autoclaves can find their way into the condensing droplets or protocells as a result of the pressure decreasing in the system (e.g., by the geyser erupting). This process took place perhaps 50 times a day in billions of autoclaves for thousands of years. It is easy to estimate that after a certain point in time, after endless "experiments" in some cases, all of the building blocks required for biological cell development found their way into one of these vesicles at the same time. The RNA with all stored information about the requisite components and the P-ribosome was necessary for all, in addition to the tRNAs, synthetases, and enzymes, which proved to be helpful in copying the RNA. Enzymes that formed ion channels in the cell membrane were particularly important for ensuring mass transport from the outside in and the inside out. Other enzymes that controlled the production of the building blocks for the cell membrane were also of great importance for the further development. All these building blocks were required immediately.

Iron-Sulfur Cluster

The formation of iron-sulfur clusters (Fe-S clusters), which are linked to structured peptides through cysteine, an amino acid with a built-in sulfur atom, played an early role in the development of the first cell. Fe-S clusters are believed to be the oldest naturally occurring biocatalysts. A common form is the cube-shaped cluster with four iron and four sulfur atoms each. It is reminiscent of pyrite, the mineral that Günter Wächtershäuser attached such great importance to in the discussion on biogenesis. Fe-S proteins are involved in many key cell processes, such as photosynthesis and the biosynthesis of amino acids, nucleic acids, and proteins. This demonstrates great importance for the viability of the cells and is an indication that the first clusters must have been involved in the development very early on. In connection with complex enzymes, a large number of functional molecules have developed from them in the course of evolution. Through the ability of iron to both donate and adopt electrons (Fe^2, Fe^3, reduction and oxidation), they probably first played an important role in processes where redox processes played a role. Pyrite (FeS_2) is a mineral that can crystallize in many places under oxygen-free conditions, including from hydrothermal solutions in crevices. The starting materials can be present in a wide range of concentration. Corresponding clusters can be integrated into proteins in a test tube under oxygen-free conditions with higher concentrations of iron and sulfur. Supporting enzymes are not required for this [7, 8]. Concentrations of iron and sulfur ions of this magnitude would, however, be fatal for cells in free form today. However, other requirements existed for the open system of micro-autoclaves in the fault zones. This once again shows the advantage of molecular development outside of a cell compartment, in which completely different substance conversions and concentrations can be assumed.

The first successful act of placing all the necessary molecules in a vesicle made all the requisite building blocks for further development immediately available and in large excess, as well, since significantly more molecules could be taken up than was necessary. The adequate supply of requisite building blocks from the outside, e.g., through ion channels, provided all the conditions for the multiplication of all parts in a narrow space from this time point onward. Ammonia (NH_3), hydrogen (H_2), carbon monoxide (CO) and carbon dioxide (CO_2), phosphate (H_3PO_4), metallic cations, and organic molecules are available.

The conditions inside the vesicle/cell led to the constant duplication of all the components, including the molecules that make up the cell membrane, so that it could grow as a whole. From a certain size, a slight elongation and low shear forces from turbulence in the water were enough for the wall to meet in the middle and form a bridge. We are familiar with this image in large from pedestrian precincts, where street performers sometimes create huge soap bubbles for the passing public for a few cents. Light winds quickly deform these up to 1-m-sized bubbles stretching them like sausages and making their edges meet in the middle. Sometimes they separate and form two separate bubbles. We can also assume that the first cells divided into two units of approximately the same size in a similar way (Fig. 8.7). The range of pre-duplicated components was so large that all the essential molecules were present in large numbers in both cells. No controlling enzymes existed to steer production in an efficient way. The process of molecular delivery, replication, and growth continued in both cells, so that after a short while each cell could divide again in the same way. LUCA was born and had started to multiply (although there is no way to differentiate between parent and sibling cells after the first division). This image demonstrates that a different number of copied cell building blocks already existed in each of the two cells in the start-up phase and that the composition of the cell was therefore different from cell to cell. It was the beginning of competitive development from the start. In addition, parallel to the first division, the initial process of placing suitable components in a vesicle could take place time and time again in large numbers. This means that the chance of a new start for additional cells persisted for a long time. Each new, dividing cell had a slightly different composition compared to the first, just as the sibling cells of LUCA differed. But the information structure, documented by the RNA, was the same in the entire system and therefore in all the cells that formed. It is not clear from this perspective that exactly one first cell existed from which all others in all domains stem. In accordance with this, it is more likely that a group, i.e., a large number of similar cells, developed in parallel and that an exchange of components perhaps occurred them. Common to all was the place of origin and the information store formed in the open system.

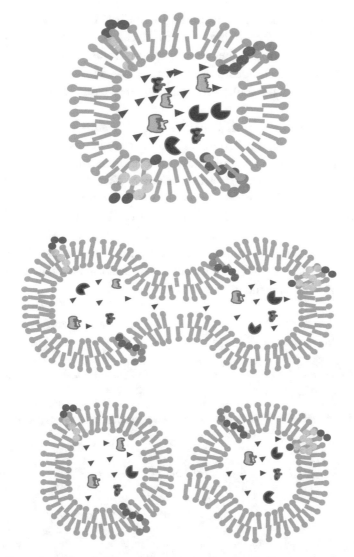

Fig. 8.7 Example of the division of a vesicle by shearing in flow turbulence after all the requisite molecular building blocks have taken their place, which can duplicate molecules in the vesicle, proteins in the cell membrane, and all other requisite building blocks. Owing to the multiple oversupply of requisite molecules, enough building blocks are available in each sub-cell even after the division, so that the process of molecular multiplication can continue to take place in both

From this point in time, with the limited development within a compartment, the supply of building blocks was limited compared to before, although the cell system was already functional. As a result, the development of new synthetases must have been subject to different laws. For example, a lot of information about peptides was stored in the RNA that had no function. Mutations in the corresponding sections of the RNA and later the DNA owing to the significantly higher concentrations of radioactive elements from this time affected the composition of the peptides. From now on, this represents an important possibility for change that could lead to functional enzymes.

The model described in this chapter is greatly simplified and excludes a large number of detailed questions that have yet to be researched. The continental crust model with its familiar framework offers enough possibilities for this. Much more involved molecular partners need to be expected, which increases the range of variation and the time required for experimentation and combination. However, it does show that solutions exist to the much-discussed chicken and egg problem. In what follows, the gap needs to be closed which lies between the formation of the initial molecules with the selection processes through the cyclic vesicle formation (Phase II) and the postulated start of life (Phase V) with the specific loading of tRNAs.

Dissipative Systems

At first glance, life appears to breach all thermodynamic laws. Organisms produce differences in concentration and temperature and build ordered structures contrary to the increase in entropy that is required physicochemically. In these structures, entropy decreases locally, which is only possible via an exchange with the environment. This takes place, for example, by giving off heat to the surrounding system, which compensates for or exceeds the size of the decrease in entropy in the structure, thus making the total entropy after the exchange higher than before. In order to maintain the process of structure formation (i.e., the processes of life), the constant consumption of energy is required, something that can only be guaranteed in an open system. Closed systems strive for a balance in which the development of life is not conceivable.

The flow of energy in a system prevents the formation of an equilibrium state. This creates disorder and chaos. The fascinating thing is that inside such chaotic systems, new structure types can originate spontaneously, which, in turn, have order and stable structures. These were described by the Russian-Belgian physicochemist, Ilya Prigogine (Nobel Prize in Chemistry 1977), as dissipative structures. A convincing example of this is the earth's surface with its atmosphere. They form an open, energy-exchanging (dissipative) system that is far from being in equilibrium. The earth absorbs energy in the form of solar radiation and radiates heat into space. Within this chaotic system, dissipative structures, such as clouds, cyclones, or even rivers, form. Prigogine speaks of a dissipative chaos to which he attributes a key role. He classifies it as being between pure chance and redundant order and evaluates it as a condition for the creation of information in biological systems [9]. Thermodynamically, the fracture zones in the continental crust, through which gases flow, form an open system with energy and material supply and an entropy flow, which is particularly evident at the transition from supercritical to subcritical gas. It comprises a dissipative system, in which dissipative structures, such as vesicles, peptides, or RNA, can occur. The formation of the building blocks for life in the necessary quality and quantity is inconceivable in a closed system. This means that if the development of RNA and proteins had been restricted primarily to a cell compartment, the cell membrane would have had to guarantee an open system from the start, as we know it today with the incorporation of membrane proteins into the cell envelope. Phase VI of the hypothetical model shows the first cell with a multiple of all functional molecules that can contribute to the multiplication of the vesicle. It documents the transition from the open system, in which there is a high degree of free movement of energy, entropy, and matter, to a narrowly limited compartment. It quickly becomes clear that the beginning of life, the first successful division process that established the development of life, could only take place with membrane molecules that received the cell compartment as an open system.

8.6 Phase III: Loans to the RNA World

Following the chapter on the selection processes for the molecules in Phase II, we made an enormous leap and left out key Phases III and IV addressing the beginning of information storage in order to make up for it in this chapter and the one that follows. It was important to recognize in advance the continued development up to the formation of LUCA in order to gain an understanding of the special features of intermediate Phases III and IV. They describe the most important intermediate steps that created the conditions for the exact loading of the tRNA. A key intermediate step is represented by the formation of an RNA, which is the first step.

In the chapter on the different models of the RNA world, I made it clear that a randomly formed RNA is not able to catalyze enzymes that simultaneously form up to 20 different synthetases. Various previously unknown intermediate steps are required for this. I described the ribosome as a large functional molecule that organizes linking the amino acids to peptides. Under certain circumstances, the ribosome's RNA can reproduce itself and at the same time act catalytically [10]. This is put forward by the representatives of the RNA world as the most important argument for the beginning of life in connection with RNA development. This requires the availability of RNA building blocks and "energy-carrying" molecules, which are needed to assemble the building blocks into a strand. Put simply, the high-energy molecules are sprouting parts of the RNA made from base, sugar, and phosphate, to which two other phosphates are attached. Today, these include adenosine

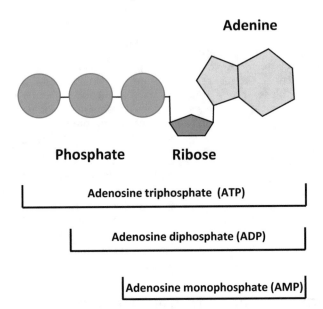

Fig. 8.8 Example of an energy-carrying molecule adenosine triphosphate (ATP). This is an RNA building block from the base adenine and the sugar ribose, to which a total of three phosphates are linked. The dividing off of one or two phosphates provides energy that is available in the cell to link other molecules

triphosphate (ATP, Fig. 8.8) and guanosine triphosphate (GTP) and in rare cases also triphosphates from the other bases. Two more phosphates (BSPPP) are attached to the phosphate on the nucleotide (base, sugar, phosphate, BSP). Furthermore, conditions must be in place that, on the one hand, allow a minimum length of an RNA strand and, on the other, prevent its rapid decay.

In our laboratories in Essen, tests in autoclaves under high pressure with supercritical CO_2 and subcritical water demonstrate initial success in linking bases with sugar (ribose) and phosphate (analyses by Prof. O. J. Schmitz, Applied Analytical Chemistry). Formation of the strand itself is about to be explored. The conditions in the crust prove to be favorable for the RNA's stability. Investigations by Järvinen et al. [11] show that at a temperature of 90 °C, the RNA's greatest stability lies between pH 4 and pH 5, exactly the range that is plausible at 1000 m depth with an excess of CO_2 and N_2 concentrations.

In contrast, in alkaline waters from "white smokers" with pH values greater than 9, RNA does not survive for long or hardly has a chance of forming. The situation is similar with the pools that dry up, where the pH values are normally over 7. The situation is different when it comes to the stability of the DNA, however. It has its optimum at higher pH values in the neutral range.

This could be due to the transition from RNA to DNA as storage developing later. All products that formed in the depths were transported to higher levels by the rising water and catapulted directly to the surface by geysers erupting. As a result, the first cells also reached higher zones in which higher pH values may have been present depending on the conditions. This was caused by surface water with higher pH values entering, which increased the pH values of the mixed water overall.

When temperatures fall below 90 °C, as is to be assumed at the depth that interests us, the RNA becomes even more stable. Taking into account the higher crust temperatures in the earth's infant days and the temperature fluctuations owing to the phase transitions for the gases, a temperature window of approx. 30 ± 10 to 70 ± 10 °C at depth of 1000 m can be estimated. This means that the RNA has optimum stability at the location where its origin is postulated, which means that longer strands can survive for a longer period of time. Its stability is further promoted by the appearance of magnesium ions (Mg^2) [12] and boron. Magnesium is provided in the crust by olivine dissolving. Olivine is one of the main minerals in basalt or gabbro rock and is found in all the deeper areas in the fault zones. Gabbros are related to basalts chemically but are equipped with larger crystals owing to their slower cooling and crystallization at depth. The hydrothermal solutions easily decompose olivine and other minerals that contain magnesium, so that magnesium ions are available in large numbers. Boron is important for the stability of ribose, the sugar used in RNA [13]. In conjunction with other ions, it forms various minerals, which predominantly crystallize in salted basins under arid conditions. In minerals that crystallize from magma, in contrast, it is only to be found to a very small extent. Boron is highly soluble in aqueous fluids and can therefore be found in an enriched form in residual solutions from crystallizing magmas and hydrothermal waters. The concentrations of the different ions in the fluids in fault zones vary with the strength of the rising gas or the inflow of water from the deep and the proportion of water that ingresses from the earth's surface. In any event, we can assume that milieus existed in the continental fault zones where sufficient boron was available as a stabilizing element for the RNA.

An intense discussion exists concerning whether the prebiotic world provided opportunities to support the formation of an RNA world. In relation to this, Szostak [10] compiled eight problem points which, in his view, can be solved with the help of new ideas and research methods. An important point here is the structural behavior of longer RNA strands. They tend to fold over, assemble in layers at suitable points, and form double strands. In these sections, they are no longer available as a template for a copy and need to be separated first for this purpose. However, the bonds are so stable that higher temperatures of more than 60° C are required [14] for separation. This temperature environment is harmful to some lipids, however, some of which are believed to be the first protocell building blocks.

As described above, larger temperature differences appeared in the continental fault zones owing to cold water geysers during the cyclical transitions from supercritical to subcritical gas. The sudden formation of gas leads to an expansion and accompanying decrease in temperature of >20 °C. The reverse situation occurs with the renewed pressure buildup owing to the water converging in the protruding water column, which leads to an increase in temperature owing to compressed gases. For a short time, temperatures are reached that are well above the average of 40 to 50 °C. Hereby the temperatures of over 60 °C determined in the laboratory for RNA replication are reached under realistic conditions, and this with each cycle determined by the geyser system. At the same time, the vesicles investigated in the Essen research group are still stable in a temperature range well above 60 °C [15].

In various laboratories around the world, experiments are performed in which Mg^{2+} ions and Zn^{2+} ions are meant to serve as supporting factors for the formation of RNA matrices (templates). The high concentrations of, for example, magnesium required for this cause interfere elsewhere. Both high and low magnesium concentrations can be found in the waters in the fault zones. They are absent in supercritical CO_2, in which inorganic substances are not dissolved. Organic bases are weakly hydrophobic, meaning that they can enrich themselves in the supercritical CO_2. The mist droplets formed when the pressure drops form a reaction space that allows a wide variety of variations for possible compounds. In the combination of the supercritical CO_2 and the water in the micro-autoclaves, the cyclical temperature fluctuations and the pH values that are optimal for the stability of the RNA mean that all the conditions for the start of an RNA world in the continental crust must have been met by a constant molecular supply of the requisite building blocks and stabilizing ions. One of the tasks over the next few years will involve confirming this RNA world in the laboratory under the simulated crust conditions.

At the moment, it represents an evidence-based assumption for the intermediate step in Phase III of the hypothetical model. Primarily, this includes an RNA being able to form in an open system, and secondly, that conditions were present making its reproduction possible. In the highly dynamic interference systems with a high concentration of radioactive elements, a high rate of change can be assumed when the first copying processes took place. This represents the crucial precondition for the model because it explains the prototype for a transport RNA, the variants of which brought the first amino acids together to form enzymes.

At this point, it becomes clear that the conditions for the formation of an RNA in the continental crust were highly favorable in terms of stability and starting materials. The development of an RNA could take place parallel to the vesicle formation and the chemical evolution of peptides, which made a large number of feedback effects possible.

In summary, it can be stated that:

- The hydrothermal chemistry in the fault zones provides the starting materials for the formation of nucleotides that lead to RNA strands being able to be linked in the $scCO_2/gCO_2$ transition zone.
- The pH values for the aqueous environment and magnesium and boron ions offer optimal conditions for the stability of the RNA.
- A double-strand formation of the RNA, which prevents the formation of copies, is resolved by cyclical increases in temperature following geyser eruptions. Subsequently, the addition of complementary nucleotides and therefore the copying process become possible. These processes take place particularly effectively in the vicinity of mist droplets in the gas phase for CO_2.
- High levels of radionuclides lead to a multitude of mutations due to ionizing radiation, which creates new RNA variants.

According to Christian Mayer, the chemical evolution of the peptides, which took place in connection with vesicle formation, could also lead to the formation of high-energy molecules, such as adenosine triphosphate (ATP) or guanosine triphosphate (GTP). These represent the two main sources of energy for the conversion of substance in the cells. Today, individual RNA building blocks (e.g., the nucleotide adenosine monophosphate) are loaded with additional phosphate molecules with the aid of a very complex enzyme system via concentration differences in load carriers (H^+) inside and outside a cell membrane. This creates high-energy molecules that are freed elsewhere from one or two phosphate molecules again under an increase in energy. The loading process is complex; the aforementioned enzyme system is comparable to a mechanical turbine that is driven by the H^+ gradients in the cell membrane [16]. This enzyme did not yet exist in the initial phase, but an infinite number of peptides developed that took their place in the vesicle membrane, some of which formed ion channels. A trans-membrane passage of ions could have been facilitated by these which in turn could have led to the formation of high-energy forms of the peptides.

Taking C. Mayer's considerations as a basis, experiments are planned in which the chemical evolution based on vesicle formation is to be coupled with the mechanisms in RNA formation. It remains to be seen whether adenosine triphosphate (ATP) or other triphosphates could have formed as energy sources that may have made a decisive contribution to the development of life. The availability of, for example, ATP in the early stages of the development of life would have given the go-ahead for the efficient linking of nucleotides to RNA strands. With the correct selection of possible inorganic and/or organic compounds, we assume that in the experiments, a strand length can be formed that allows for ribosomal properties. From this point onward, we can clarify whether other RNA molecules, such as tRNA, can also be catalyzed in addition to the copy.

The formation of the vesicles would thus retain a multiple share in the development of the life process: with the selection of special peptides/proteins through the properties of the vesicle membranes, with the creation of ATP/GTP as the most important energy building blocks, and later finally with the provision of the cell itself started from LUCA. Cyclical pressure fluctuations causing the disintegration of the vesicles led to the release of the peptides and ATP molecules with each phase transition, which could then continue to react in the open system created by the micro-autoclaves. The vesicle formation system therefore represents one of the most important facilities for producing the molecular building blocks needed on the path to the origin of life.

8.7 Phase IV: Closing the Gap

With the provision and selection of long peptides/proteins by the membranes in the vesicles, similar proteins were formed repeatedly, which belonged to different groups depending on the vesicle membrane, the amino acids available, and the physicochemical framework. In addition, longer amino acid chains also existed, which were formed independently in water and $scCO_2$ or their interfaces. Available were 10 to 12 amino acid species found in hydrothermal systems. However, a clear accumulation of the most abundant species of glycine and alanine must have taken place in the chain formation. In the acidic environment of the crustal fluids, the simultaneous incorporation of L and D configurations in the amino acids was suppressed, so that isolated enantiomerically pure proteins could form. They were folded and some were given catalytic properties. Throughout the entire process, the proportion of proteins with L-handedness was just as common as that with D-handedness.

At this point in time, there was no definition of the L configuration of the amino acids as we know it today.

This already relatively clear picture of the development of the proteins is contrasted by something still vague, which relates to the meaning and possibilities of the RNA at that time. This is because experiments can only be carried out in the future with the general conditions that are now well known to us, which can prove individual steps in the development of the RNA world in hydrothermal fault zones. Internationally, partial results on the RNA world have already been achieved in the past decades, which I will now apply below as a basis for further considerations.

To close the gap, for the further development in the cavities, the formation of a self-replicating RNA is assumed parallel to the formation of the vesicles and proteins, as described for the RNA world. The first strands of RNA did not yet contain any information that could be used to form peptides. Only its own configuration was called up with each copying process. From the large number of copies under high radioactive influence in the early days of the earth, an abundance of variations can be expected, which led to ever new combinations.

This is where the model comes into play. This envisages that, from a certain point in time, an RNA was modified in such a way that the two ends partly consisted of complementary bases so that they could combine to form a double strand. The mismatched parts were in the middle, which bent this section into a kind of loop. The contiguous ends were not of the same length, with three nucleotides protruding beyond the double strand like a spur. Similar structures for RNA molecules can now be easily created in the laboratory and are not unusual. Ultimately, the modification of the RNA described gave rise to the basic structure of a molecule, which in the further development took on the function of the transport RNA (tRNA). The possibility existed of linking an amino acid to the spur, and three bases existed in the strongest bend of the loop that were suitable as an information block for reasons of space. The double strand could be separated by cyclical temperature increases up to the melting temperature and repeatedly copied thereby. This resulted in mutations that mainly affected the area without any double-strand formation. The spur no longer changed.

The tRNA formed in this manner represented the key molecule par excellence. It could interact with enzymes from the vesicle formation and be loaded with an amino acid. At the same time, a chemical module existed at the other end that could be used as a code. At this point, it is tempting to make a comparison based on technology. We all know about industry standards and how useful they are. They make it possible for companies, for example, to use regionally manufactured devices everywhere due to standardized connections. If we look at the tRNA molecule, the idea appears to be extremely old. The tRNA molecule is one of the oldest functional molecules. From the beginning, there must have been a contact point for the absorption of an amino acid, the spur that protrudes beyond the area of the double strand. This is formed by a nucleotide sequence with the bases ACC (adenine, cytosine, cytosine). The invention of this "ACC connection module" was so fundamental and successful that it has not been changed to this day. All biochemical processes linked to the formation of peptides and proteins have taken place over this module in every single cell from the beginning on until today. It is like a standardized trailer hitch on a tractor to which all types of trailers can be coupled.

It is now a matter of identifying a process that is the reason for the selection of a specific amino acid that is not linked to anyone but to a specific tRNA. The goal, if we follow the model for the previously presented Phase V, is to form just two synthetases from only two amino acids.

We can assume that not every tRNA was able to interact with each of the randomly formed enzymes. This is a matter of course today after billions of years of adaptation. In most cases, the coordination of associations is absolutely perfect after this long period of time. If errors occur caused by mutations that weaken the exact assignment, disadvantages usually occur along with long-term system death. More complex molecules already existed with the first enzymes formed in the vesicle cycle, but the selection of optimal tools that only accepted very specific tRNAs for the further steps was still completely missing. Of interest among the enzymes formed early on are those that have synthetase properties. That means that folding structured them in such a way that they could offer both a pocket for a specific amino acid and a docking site for an tRNA, albeit very inaccurately. Each enzyme was unique however, and after its decay, the same could not be supplied again. The variability was too great for this owing to the random formation. For each newly formed enzyme, however, a group existed in which the relevant enzymes had similar properties (hydrophobic, hydrophilic, basic reaction, etc.). And each group had different physicochemical properties, which resulted in the general preselection of the type of amino acid that could bind them electrostatically. At the same time, each group had a preference for certain tRNA molecules that offered a wide range of variations through mutations.

The formation of groups of proteins/enzymes in combination with different tRNA molecules provides the first chance to sort amino acids out from the range of molecular soups. For subsequent reactions, this means predetermining and narrowing down the possible variations. Not every tRNA can be loaded with every amino acid by every synthetase. Even if the specificity of the bonds is very weak, a total of processes results over long periods of time that produce similar reaction products. And time is something we have a lot of.

Analysis of the oldest part of the genetic code in today's cells shows that the amino acids glycine and alanine and the associated codons in the genetic alphabet most likely contributed to the start phase of life [17]. Glycine is the simplest amino acid, followed by alanine. Glycine is present in enzymes today at a proportion of 7.5%, and alanine is also represented with the highest value of all of 9%. Alanine is hydrophobic. Glycine is a little more complex, since differences in properties exist that depend on whether glycine is available as a free molecule or within a peptide. It may mean that the first links made by amino acids into longer chains were primarily made from these two representatives following preselection by the processes described above. Decisive factors may have been the properties in relation to water and the quantity of molecules.

We have arrived at this point at a completely disordered process based on random contacts by amino acids with randomly formed proteins/enzymes and randomly formed tRNAs. The tRNA has a block of three, consisting of bases (anticodon) that become an important information carrier in the further course but do not contain any useful information for targeted molecular reactions at this stage. This therefore provides us with a coupling of two systems that just results in very unspecific reactions with one another (cf. Sect. 8.3). The question now is how a specific assignment option could have developed from it that ultimately resulted in the information being stored in an RNA.

Under the starting conditions described, the possible conditions available in a micro-autoclave in the continental crust are as follows: under the special conditions in the cavities, amino acid chains formed which belonged to certain groups of peptides/proteins despite their large variation in the sequence. With the help of their membranes, the vesicles sorted out certain groups that interacted with the vesicle shells. In addition, other groups emerged separate to the vesicle contact. From a certain length, enzymes from all groups were folded, which gave them greater stability. They consisted of the amino acids available in the hydrothermal environment (about 10 species), but with a significant accumulation of glycine and alanine. Among them were enzymes that had catalytic functions and contributed to the tRNAs being loaded with amino acids (synthetases). This represented the first functional contact between the RNA world and the protein world.

Some enzymes always existed within the groups that preferred to have a pocket for the amino acid glycine and a docking place for a suitable tRNA, and some provided the same for the amino acid alanine. Since the assignments were only very weakly specific, other amino acids could also take up the places for glycine and alanine, and it was also possible for the tRNAs to be exchanged in part.

How can we imagine this taking place? How could a peptide or protein that only consisted of the two most common amino acids be formed under these general conditions?

The combination of amino acids, tRNAs, and certain groups of enzymes resulted in loaded tRNAs that flowed through the fluids in the cavities in high concentration. They arrived at RNA strands that were able to act as catalysts (P-ribosomes). The loaded tRNAs collect on them time and time again, briefly form layers together, and link the amino acids they have brought with them. These separated from the tRNAs and gradually formed a chain. Glycine and alanine must have made up most of the supply in these processes. It was only a question of time before the many variations in chain formations included some that, because of the oversupply, only consisted of glycine and alanine, which were then folded and able to form an enzyme. However, despite the low pH values in the cavities, a mixed incorporation of the L and D versions of the amino acids must have occurred. That may not have been a disadvantage since plenty of opportunities were available to try out new variations again and again. But the number of attempts needed to arrive at an enantiomerically pure peptide would have been so high that the matter would have quickly come to a dead end. The lucky coincidence that glycine is the only achiral amino acid comes into play here, so that a handedness cannot be assigned to it. That means that the combination of glycine and alanine had to be tried out until either only D-alanine or only L-alanine was present in the peptide/protein. And the probably simply structured synthetases could afford to do this with the long time periods available to them.

At this point, only one crucial step is missing for finding a solution to the chicken and egg problem. While the amino acids on the P-ribosome providing assistance are linked after the tRNA has been attached, the respective three bases of the tRNA (anticodon) are so favorable that complementary RNA building blocks can attach themselves. If they are linked, they form a codon that is the counterpart to the anticodon of the respective tRNA. If this process occurs for each subsequent tRNA molecule that docks onto the P-ribosome and the codons are also linked, a new RNA is gradually created.

What have we gained until now? We hypothetically formed a chain containing only two amino acids, which demonstrate a certain specificity to tRNAs (here the number of usable tRNAs per amino acid was greater than one, as is the case with most amino acids today). At the same time, an RNA developed over the anticodons from the tRNAs, which represents the first unit of information in relation to the amino acid chain formed. (If this process were to be followed in the laboratory, at this juncture, it would be appropriate to break open a bottle of champagne!) And the conclusion? The conclusion is that the most important molecule in the overall process for life, the tRNA molecule, connects the protein world with the RNA world for the first time over a logical link.

So, where do we go from here? Well, now we can really get started. We have enough tools together to prepare for the start phase for life. First and foremost is the continuous formation of peptides/proteins, which consist of glycine and alanine and have a random sequence. Under appropriate conditions, the sequence can be stored immediately in an RNA formed in parallel (Fig. 8.9). Of the many randomly formed proteins, two have synthetase properties and can specifically load glycine or alanine onto suitable tRNAs. If the sequences of the two synthetases are stored in an RNA, survival is ensured. For this to succeed, the tRNAs have to dock onto the codons on the RNA (messenger RNA or mRNA today) and combine the amino acids in the order given therein. This could again be done most elegantly with the help of a P-ribosome.

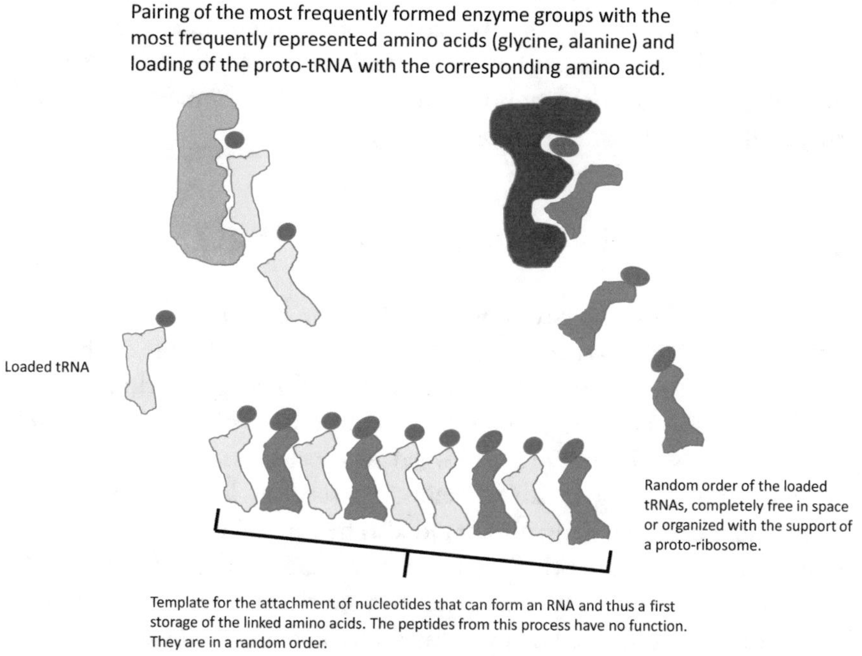

Pairing of the most frequently formed enzyme groups with the most frequently represented amino acids (glycine, alanine) and loading of the proto-tRNA with the corresponding amino acid.

Loaded tRNA

Random order of the loaded tRNAs, completely free in space or organized with the support of a proto-ribosome.

Template for the attachment of nucleotides that can form an RNA and thus a first storage of the linked amino acids. The peptides from this process have no function. They are in a random order.

Fig. 8.9 Pairing of the most frequently formed enzyme groups with the most strongly represented amino acids (glycine, alanine) and loading of the proto-tRNA with the corresponding amino acid

The consequence of the existence of two specific synthetases, which leads to the formation of peptides that can be documented in an RNA, is described in Phase V (Sect. 8.3).

The decision in favor of the L- or D-configured system of amino acids needs to be made at this point at the latest. Both systems had been running in parallel up to this point. This could only be explained by the fact that a minimally larger number of L-alanine were present in the system during this time. This small overhang then caused the chain with the L-handedness to fold into an enzyme, thus causing Phase V to start to develop. This system prevailed over time because all subsequent steps were based on the beginning. In any event, the representatives of the version that built the first two specific synthetases won. It is virtually impossible for this case to occur simultaneously for both handedness. And so, with the start of the propagation system for synthetase, the breakthrough was achieved and the race (alone in terms of resource requirements) was won.

Needless to say, this simplified process must have been far more complex, perhaps involving more amino acid species and more tRNA variations. But this cannot be discussed until the very end. However, the time factor does exist that allows the acceptance of a myriad of failed attempts and variations with probabilities that can never be reproduced in the laboratory. That said, however, the issue of concern here is the basic principle which, once understood, allows the essential key point stone in the development of life to be narrowed down and examined (see also [18, 19]).

8.8   Can It Really Have Been Like That?

In our considerations, we have now linked a large number of amino acids in cavities in the earth's crust to form peptides and subjected some of them to a relatively broad sorting process by continuously forming and decaying vesicles. The processes here contributed to a concentration of mixed amino acids in peptides (Phase I). Low pH values led to selected amino acids with only one orientation linking to form chains. As soon as longer peptides were formed, more complex structures could also be formed by folding. The next step involved a selection, which led to the formation of a peptide with few species (in the case under consideration two species) (Phase II). In parallel, a ribosomal RNA was formed that was capable of copying itself and from which tRNAs emerged. The tRNAs interacted with the continuously emerging peptides from Phase I, with the result being that they were each linked to certain amino acids (Phase III). A natural oversupply led to the preferred selection of the two simplest amino acid species, which were transported by different tRNAs and linked to new peptides with their own amino acid sequence (Phase IV). Here, the anticodons from the tRNAs could be used as the first template for a simultaneously forming RNA. After an indefinite period of time, one (or a little later two) enzyme was present, which retained the property of a tRNA synthetase and whose sequences were stored in an RNA. These must have been the type that could each catalytically bind one of the two own amino acid species built into them and connect them to one of the tRNAs. The advantage from now on was that the sequence of the amino acid species from the two chains was stored in an RNA, so that the synthetases could be reproduced continuously. The free attachment of loaded tRNAs was also possible in parallel, which led to a large number of nonfunctional peptides. In some cases, their sequences were also stored in an RNA formed in parallel. There were occasionally enzymes that were synthetase-like and able to load a third, fourth, fifth, etc. tRNA with its own amino acid. If information was stored for them in an RNA in parallel, this made it part of the equipment for the entire system.

Following the development of the minimum building blocks required for a dividing cell, the opportunity for the start had arrived. All that was required was to place the right selection of components in one of the protocells into one of the countless vesicle formation cycles. An important prerequisite for this was for the number of building blocks entering the vesicle to be significantly higher than required. Membrane proteins, which allowed the passage of "nutrients" and protons, as well as the entire machinery for enzyme formation and transfer of the information, were also important. The supply from the outside through the cell membrane meant that both the complex molecules of the inner life and the lipids of the cell envelope could be continuously reproduced. The latter led to the growth of the cell membrane, which, when reaching a critical size, was divided by simple physical shearing. The high number of all preformed molecules ensured that in both sub-cells emerging from the mother cell, more building blocks for a "basic equipment" were available than was minimally necessary, meaning that division process could be repeated in the same way a short time later for both cells. LUCA had started to multiply (Phase VI).

This makes it clear that the parallel development of peptides, vesicles, and RNA molecules outside of a cell compartment delivers a great advantage. The most important components are in an open system and are supplied by it. The complicated molecules required for the functions and processes within a cell only play a role later. These include the enzymes that facilitate mass transportation in the cell membrane and the enzymes that use gradients for generating energy or provide assistance in manufacturing the molecules for the cell membrane and cell division. These enzymes or their precursors were able to develop in the crevices of the crust as a result of the process described and were repeatedly incorporated into the cyclically generated vesicles in infinite numbers of experiments.

In Sect. 4.1, I referred to the possibility of getting information about LUCA from the genome of the closest possible relatives, such as methane-forming archaea and other prokaryotes. The investigations have shown, as would be expected, that LUCA lived anaerobically, fixed carbon dioxide and nitrogen, was dependent on hydrogen, and preferred a hot environment [20]. Iron, sulfur, molybdenum, nickel, and selenium were needed for certain proteins and compounds. Everything indicates that this comprised a hydrothermal environment in which carbon dioxide, hydrogen, nitrogen, iron, and other metals were present in sufficient numbers.

If we look from a distance at the process outlined of creating LUCA, we can again make a vivid comparison with technology. Let us consider an internal combustion engine in relation for a moment which is in a perfect state to immediately start up. But it won't. The reason is simple: to start the engine, we need a starter motor. Perhaps life was exactly the same. Perhaps all of the building blocks were in the right place at the right time, and it just needed a physical process to help it start. If we abstract the processes described in a gas-bearing fault zone a little, we can recognize a kind of steam engine that fails to work (Fig. 8.10). The inner earth takes on the role of the boiler. The super-critical gas rising from the deep converts into subcritical gas when a limiting value is exceeded and pushes the "piston" upward. In this case, the piston is not a solid material, but simply water. Work is performed when it is pushed out. The "piston" returning corresponds to the water returning, which restores the initial state. The process takes place again if the system has enough energy stored in it. The clock frequency is in the tens of minutes range, as we know from the geyser eruptions. The question is what work and energy are deployed in these processes for the chemical processes. It only represents a very small part: the generation of turbulence in the cavities and heat being conducted into the fluids. Here, too, there is no fixed connecting rod that transfers the piston movement in a rotary motion; the water in a liquid, flowing form is sufficient. The turbulence and the phase change for the gas represent the causes of many chemical reactions and the cyclical formation of the vesicles. The freely available energy reserves in the interior of the earth (geysers have existed for more than four billion years) and the extreme increase in entropy on each eruption provide optimal conditions for the reaction sequences.

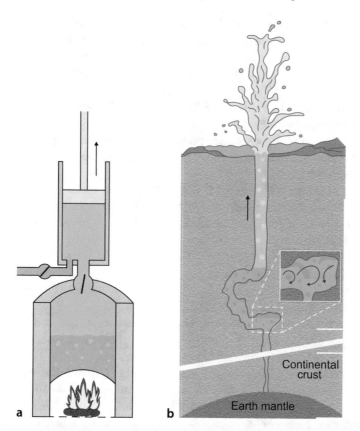

Fig. 8.10 Comparison of the simplest steam engine with a geyser system. Hot steam is fed into the cylinder of a steam engine which, supported by a counterweight, pushes the piston upward. Cold water that is sprayed in afterward reduces the volume of the steam by creating condensation. The resulting negative pressure pulls the piston back down. In the case of the geyser, the piston is made of water, which is pushed out and runs back down in the next cycle, increasing the pressure at depth

To put it bluntly, if it was thus, we are ultimately the product of a kind of steam engine. And after four billion years, this product succeeded in launching an improved version onto the market. It ushered in a new revolution that may have been similar in meaning as the origin of life for the next billion years.

References

1. Simoneit BRT (2004) Prebiotic organic synthesis under hydrothermal conditions: an overview. Adv Space Res 33:88–94
2. Moosmann B (2017) Molekulare evolution: redoxbiochemie des genetischen Codes. BIOspektrum 23(17):748–751. https://doi.org/10.1007/s12268-017-0864-7

3. Süssmuth RD, Mainz A (2017) Nonribosomal peptide synthesis—principles and prospects. Angew Chem 56(14). https://doi.org/10.1002/anie.201609079

4. Danger G, Plasson R, Pascal R (2012) Pathways for the formation and evolution of peptides in prebiotic environments. Chem Soc Rev 41:5416–5429

5. Eriani G, Delarue M, Poch O, Gangloff J, Moras D (1990) Partition of aminoacyl-tRNA synthetases into two classes based on mutually exclusive sets of conserved motifs. Nature 347:203–206

6. Delarue M (2007) An asymmetric underlying rule in the assignment of codons: possible clue to a quick early evolution of the genetic code via successive binary choices. RNA 13(2):161–169

7. Beinert H, Holm RH, Munck E (1997) Iron-sulfur clusters: nature's modular, multipurpose structures. Science 277:653–659

8. Lill R (2009) Function and biogenesis of iron-sulphur proteins. Nature 460:831–838

9. Nicolis G, Prigogine I (1977) Self-organization in nonequilibrium systems. Wiley-Interscience, New York

10. Szostak JW (2012) The eightfold path to non-enzymatic RNA replication. J Syst Chem 3(2). https://doi.org/10.1186/1759-2208-3-2

11. Järvinen P, Oivanen M, Lönnberg H (1991) Interconversion and phosphoester hydrolysis of 2′,5′- and 3′,5′-dinucleoside monophosphates: kinetics and mechanisms. J Org Chem 56:5396–5401

12. Fischer NM, Polêto MD, Steuer J, van der Spoel D (2018) Influence of Na+ and Mg2+ ions on RNA structures studied with molecular dynamics simulations. Nucleic Acids Res 46(10):4872–4882. https://doi.org/10.1093/nar/gky221

13. Furukawa Y, Horiuchi M, Kakegawa T (2013) Selective stabilization of ribose by borate. Orig Life Evol Biosph 43(4–5):353–361. https://doi.org/10.1007/s11084-013-9350-5

14. Wienken CJ, Baaske P, Duhr S, Braun D (2011) Thermophoretic melting curves quantify the conformation and stability of RNA and DNA. Nucleic Acids Res:1–10. https://doi.org/10.1093/nar/gkr035

15. Mayer C, Schreiber U, Dávila MJ, Schmitz OJ, Bronja A, Meyer M, Klein J, Meckelmann SW (2018) Molecular evolution in a peptide-vesicle system. Life 8(2):16. https://doi.org/10.3390/life8020016

16. Yoshida M, Muneyuki E, Hisabori T (2001) Atp synthase – a marvelous rotary engine of the cell. Nat Rev Mol Cell Biol 2:669–677

17. Trifonov EN (2009) The origin of the genetic code and of the earliest oligopeptides. Res Microbiol 160:481–486

18. Carter CW (2015) What RNA world? Why a peptide/RNA partnership merits renewed experimental attention. Life 5:294–320

19. Carter CW (2016) An alternative to the RNA world? Nat Hist 125:28–33

20. Weiss MC, Sousa FL et al (2016) The physiology and habitat of the last universal common ancestor. Nat Microbiol 1:16116. https://doi.org/10.1038/NMICROBIOL.2016.116

9

Life = Order + Complexity

Abstract We know that life requires a high degree of structural order. However, even extremely ordered structures like crystals have no tendency to be alive. Likewise, it is accepted that life is based on a high degree of complexity. However, even very complex systems like mixtures of numerous organic compounds formed by uncontrolled reactions just remain dead matter. So neither order nor complexity alone can guarantee for life. In order to cross the limits toward life, a system has to fulfill both conditions simultaneously. Order and complexity together turn out to be an essential pair of characteristics for life. It actually has the power to define life and its evolution, as well as to explain the principle for the first steps of prebiotic chemistry.

9.1 Searching for Life

Suppose you were an astronaut on a foreign planet, searching for life. What would you actually look for? Of course you have to consider that unknown life may appear in a form which is completely different from all life we know. It may look and function so differently that any comparison with terrestrial forms of life would be impossible. So in your search for life, it does not make sense to look for bushes or trees with green leaves and for crawling insects or spiders. Instead, you have to find markers which reflect the basic principles of life: structural features or processes which indicate "a self-sustaining chemical system capable of Darwinian evolution" which is the NASA's definition for life [1]. But how to recognize them?

© Springer Nature Switzerland AG 2020
U. C. Schreiber, C. Mayer, *The First Cell*, https://doi.org/10.1007/978-3-030-45381-7_9

What is the basic principle of life? In 1945, during his exile in Ireland, the Nobel laureate physicist Erwin Schrödinger wrote a book titled *What is Life?* [2]. In this book, he states that "life decreases or maintains its entropy by feeding on negative entropy." If we roughly translate "entropy" into "disorder" (see Sect. 3.5), this statement would mean that life is an ordered system which maintains or increases its order by feeding on negative disorder. Schrödinger seems to identify the order as a central parameter for life. So maybe, wandering on the planetary surface, you should look for ordered structures? We know that entropy is a measure for disorder, so looking for order means looking for low entropy. On your field trip to a foreign planet, you of course enjoy the latest and best analytical equipment, so we assume that you carry a portable entropy meter (a device that still needs to be invented). So you stumble over the rugged planetary surface and point your instrument here and there and watch out for low entropy values. Soon you will notice that the readings will be close to zero whenever you find crystalline minerals. Structures like quartz crystals, diamonds, or pieces of salt represent nearly perfect order. With the atoms or ions placed in a regular three-dimensional lattice, they are among the most ordered structures in universe. But are they alive? Of course not. For being alive, they lack the extremely diverse functional units of machinery which are able to undergo metabolism, growth, or reproduction. In other words, they lack complexity. So obviously, order alone is not a sufficient criterion.

So the next day, you try another excursion, this time equipped with a complexity meter (again a device which is still dearly missed in today's laboratory equipment). Pointing that instrument onto the minerals you found the day before, the needle will hardly move: the building principle of a mineral crystal is generally very simple as it just involves endless repetitions of small elementary units, the so-called elementary cells of the crystalline lattice. So you keep moving across the surface until you suddenly detect high complexity in an amorphous black chunk of material. It was formed from basic organic compounds (which are commonly found on interplanetary bodies such as comets) under the influence of UV radiation over extended periods of time. Under these circumstances, small organic molecules can form chains of variable composition which by themselves can undergo further reactions. Altogether, these reactions lead to extremely complex chemical compositions. In its touch and in its optical appearance, the mixture could resemble common asphalt. Is it alive? Of course not. Even if the material may compete with an organism regarding its chemical complexity, it lacks the structural regularity which is required for the complex functionality of life. In other words, it lacks order.

Now you are at a crucial point. You realize that, in order to be alive, a system has to combine order and complexity. It is exactly this combination which separates dead objects like crystals or asphalt from something like a living bacterial cell. In such a cell, we will find an ordered subcellular structure with membranes, separated internal environments, concentration gradients, and large molecules with well-defined three-dimensional structures. At the same time, the cell is very complex, with thousands of chemical constituents, numerous sets of interdigitated chemical reaction schemes, and a complex chemical memory stored as a genetic code. The order inside the cell and the structural complexity of the cell form a functional unit; one cannot work without the other. So, for a successful search for life, you will have to carry two instruments since you will have to determine two important values simultaneously: order together with complexity. With this in mind, one may try to define the limits of life. Where does it start? What are the limits regarding order and complexity? How can you assign measurable quantities to the order and the complexity of life?

9.2   Life and Order

In terms of order, the crucial threshold value is connected to the entropy which is abbreviated by the symbol "S" (in units of Joule per Kelvin, or J/K). Since entropy roughly measures disorder, we may introduce reciprocal entropy 1/S as a measure for order (now measured in K/J). Let us look at a living cell as we know it from terrestrial life. Here, the most important contribution to order (or to 1/S) consists in the fact that most chemical constituents of the cell are not distributed equally but occur at specific positions of the cell in subcellular structures, separated by internal membranes. This particular contribution to order is absolutely essential to keep the cell alive: the concentration gradients across membranes are connected to important functions and energy sources of the cell. Actually a common test on the viability of the cell, a so-called life dead assay, relies on the fact that the membranes of dead cells become permeable [3]. So it is actually this membrane-induced order which separates a living cell from its death. This critical difference can be determined as the mixing entropy which occurs on homogenization of all cell contents. In other words, the deadly loss of order can be measured as a fraction of the disorder which would be induced if all cell contents would be scrambled by something like a microscopic mixer. So taking the reciprocal entropy of a homogenized cell as a reference point, we have a very useful scale to determine

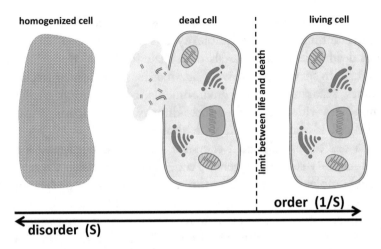

homogenized cell   dead cell   living cell

limit between life and death

order (1/S)

disorder (S)

Fig. 9.1 Different states of a biological cell on a scale of order and disorder. Even the rupture of a single membrane can cause the death of the cell which then crosses the line between life and death (right). A reference value for order is given by the state of a completely homogenized cell (left)

the level of the reciprocal entropy which, based on the form of life we know, separates life from death (Fig. 9.1).

So far we have been looking at static aspects of life. What about processes? Within a living cell, numerous chemical reactions occur simultaneously which are connected to life. These reactions form a complex and well-ordered network with one reaction using the products of another. All these reactions occur in a controlled manner such that accumulations or shortages of reaction components are avoided. How is this reaction network accounted for in terms of order?

We know that every process in universe tends to increase disorder (see Sect. 3.5). Consequently, every chemical process inside the cell basically would have the tendency to drive the state of the cell toward the line of death indicated in Fig. 9.1, and eventually it would kill the cell. So how can the cell survive any process over an extended period of time? How can the cell avoid the seemingly unavoidable death of disorder? The answer was given by Erwin Schrödinger in a somewhat cryptic statement: the cell has to "feed on negative entropy" [2]. But what does that actually mean?

In fact, life has invented the solution to this problem at a very early stage: energy. Life uses energy to create positive entropy, or in other words, disorder in the environment. Chemical energy, as in food, turns into heat, and heat turns into chaotic motion of molecules. However, a fraction of this disorder is taken back, as negative entropy. The chaos induced by burning one gram of sugar, for example, is slightly larger in an open fire than in case of a biological degradation process in a living cell. The difference is exactly the negative entropy the cell is feeding on. With that portion of negative entropy, the cell manages to stay at a constant position of the scale of order and disorder (Fig. 9.1) and allows it to run continuous reaction networks. With the use of energy to create a portion of negative entropy, life has discovered a way to prevent the death of disorder while keeping up the dynamics of metabolic processes.

9.3   Life and Complexity

Regarding complexity, the approach for a quantitative value is much more demanding as many definitions of complexity exist [4]. A very suitable approach to measure complexity was found by the Russian mathematician Andrey Nikolaevich Kolmogorov. According to him, the complexity of a system is determined by the minimal size of a computer program (in bytes) which is necessary to fully describe the system in all its details. If we consider, for example, a crystal with atoms at well-defined positions in space, such a program would be very simple. It would contain the basic building principle of the crystal, the so-called elementary cell, and would just repeat this scheme over and over again. Such a program would just need a few bytes. According to Kolmogorov, the complexity of such a crystal would be very low. Now if we look at the chunk of asphalt on our foreign planet, the situation is very different. With its millions of compounds, many of them chain molecules with various side groups and differing sequences of structural units, a corresponding computer program for its complete description would be quite lengthy: every single compound, practically every single molecular chain, would have to be determined in its full structural detail. With that, the corresponding program code would become very long, easily in the range of megabytes. So, according to Kolmogorov, the complexity of this chunk of asphalt would be extremely high. But how could we define the complexity of life?

In case of life, there is in fact an existing analogy to a computer program. It is the genome, which is actually interpreted and translated into complex molecules such as enzymes and structural proteins. These in turn are responsible for the formation of cell structures as well as for guiding the complex reaction network of the cell's chemistry. That means that all the complexity we can assign to a living cell corresponds to the complexity of the cell's genome.

Very similar to a computer program, the information content of the genome can be measured in bits and bytes. In computer technology, a byte measures the information of an eight-digit binary number, which accounts for numbers between 0 and 127. With the DNA strand consisting of four different bases (forming base pairs together with their counterparts, see Sect. 4.2), the same amount of information can be stored in a section of the DNA which is four base pairs long. So the information content of the DNA can be measured in bytes, which, following Kolmogorov's definition, also measure the initial complexity of the corresponding organism.

It is surprising how little of information is necessary to code simple life. For example, a very simple microorganism called *Nanoarchaeum equitans* can exist on a genome with only 490,885 base pairs [5]. This corresponds to approximately 123 kbytes, according to the estimation given above. If we consider that a printed page like this one contains something like 2 kbytes of data, this microorganism can live on less than 62 printed pages of information. This is just two or three chapters of the book you are holding in your hands! And there may be even smaller genomes: the lower limit of genome size for living cells has been estimated to be 375,000 base pairs corresponding to approximately 93 kbytes [6]. The human genome, on the other hand, consists of about three billion base pairs with about 809 Mbyte of information. However, only something like 1.5% of this information is actually translated into protein structures, while the vast majority is reserved for regulation of gene expression and for chromosome architecture or may not serve any purpose at all [7]. Therefore, the true value for complexity of the human genome can only be roughly estimated; it may amount to something like 12 to 800 Mbytes. Compared to the genome of the microorganism *Nanoarchaeum equitans*, the complexity of a human's genome is at least a hundred times larger. And, in contrast to a simple microorganism, a human accumulates further complexity when growing up. Memory, experience, learning, and growing intelligence add a lot to the original complexity over years of life.

9.4   The Limits of Life

With these ideas about order and complexity of life in mind, your search on the surface of the foreign planet will become more substantial. For each object which you analyze, you determine both the order as reciprocal entropy and the system complexity according to Kolmogorov. Whenever a system exhibits a high degree of complexity, every process will rapidly tend to decrease its structural order. Only the mechanisms of life, the process of "feeding on negative entropy," can maintain order and create new ordered structures. Especially a system with complex processes cannot remain in an ordered state if it is not alive. Therefore, if such a highly complex system is highly ordered at the same time, it must be either some form of life itself or a structure which life has created. Whenever both values show high readings, you may have discovered either a form of life or its products or remnants. This in mind, one may create a scheme which helps to define the limits of common biological life according to order and complexity [8]. In Fig. 9.2, a diagram is shown where system complexity is plotted against structural order.

In such a diagram, the results of the first unsuccessful results of your search are easily localized. Crystals exhibit an extremely high degree of order but very little complexity; therefore they are found near the lower right corner of the diagram. On the other hand, the piece of asphalt is extremely high in complexity but very low in order, so it is located near the upper left corner.

The lower limit of life regarding order is defined by the reciprocal entropy of a living cell (vertical broken line in Fig. 9.2). It is this line which is actually crossed when the cell dies, e.g., caused by chemical or mechanical influences.

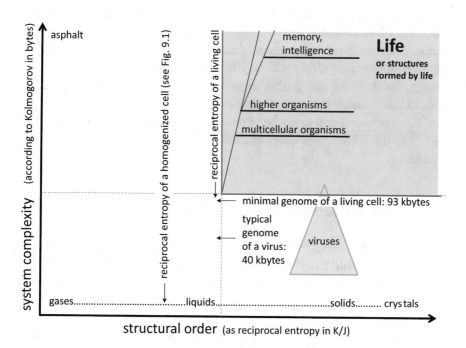

Fig. 9.2 Diagram showing the limits of common biological life according to the criteria of order and complexity [8]. The blue rectangle can be seen as the area of life limited by the minimal order (or reciprocal entropy) and the minimal complexity (or the size of the minimal genome) of a living cell. Everything which falls into this area is either life itself or any product of life like books, software, artificial intelligence, art, or culture. The scaling of this graph is arbitrary

So based on the knowledge we have on present life on earth, this limit is sharp and well defined.

The lower limit of life regarding complexity is somewhat blurry. As mentioned above, the lower limit of information for very primitive cells may be around 93 kbytes [6]. However, the corresponding organisms tend to depend on biomolecules which are present in their environment and sometimes are suspected to be parasites, such as *Nanoarchaeum equitans* [5]. So it is a question where real independent life may start. Also, there are biological structures which generally are not considered to be alive: viruses. A typical virus carries a genome of about 40 kbytes in size, which is well below the limit of 93 kbytes. However, there are also viruses with a genome as large as 140 kbytes, which would exceed the complexity value for primitive cells. On the other hand, these so-called pandoraviruses are a special case and have even been discussed as a special domain of life [9]. All in all, there is some justification to estimate the lower limit of complexity near 93 kbytes (horizontal broken line in Fig. 9.2).

With these boundaries, we may define a specific area of life. All known structures which, according to biologists, are alive on earth will fall into this section. On the other hand, it seems that every single structure on earth which falls into this section either will be life itself or was created by life in some manner. Products of life in this sense are, e.g., books, computer software, artificial intelligence, pieces of art, the Internet, or science and culture in general. These products can easily be as complex and as ordered as their creators. They may in some cases even exceed their state in terms of order and complexity.

9.5   Consequences for the Origin of Life

What does all that mean for the origin of life? Looking at our planet in a very early stage, we would have found only very simple chemistry, small molecules like water, carbon dioxide, methane, or ammonia. In our diagram, we would locate that stage at the lower left corner, as this early state would be very low in order and complexity (Fig. 9.3).

From there, a development toward life means that we have to follow a diagonal line (blue arrow in Fig. 9.3) which in fact is a problem. It is quite easy to advance in a horizontal line which means a mere increase of order: a simple crystallization process by drying or cooling a mixture of chemical components will do this job. It is also quite simple to follow a vertical line, the mere increase of complexity: this could be the consequence of numerous

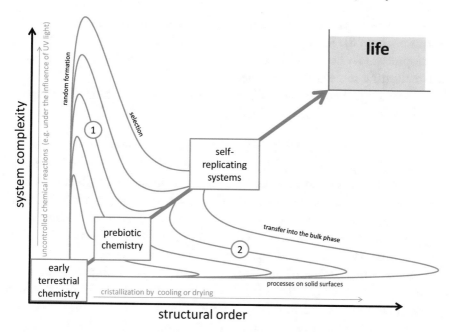

Fig. 9.3 Diagram showing the possible pathways toward life according to the criteria of order and complexity [8]. Starting with simple terrestrial chemistry (water, carbon dioxide, methane, etc.), the development went toward early prebiotic chemistry (amino acids, lipids, sugars, etc.). At some point, self-replicating systems were formed which then could undergo Darwinian evolution eventually leading to biological life. Green arrows mark spontaneous processes which are easily achieved. Pathways 1 and 2 are explained in the text. The scaling of this graph is arbitrary

uncontrolled chemical reactions, e.g., induced by ultraviolet light. But it is in fact very difficult to simultaneously increase order and complexity, that is, to advance along the diagonal of this graph.

The most powerful approach for the diagonal pathway is in fact Darwinian evolution. It proceeds in very small steps of small random changes followed by selection which in our diagram would appear as a narrow zigzag, hardly deviating from a diagonal straight line increasing order and complexity. On this path, evolution has led from single cells to higher organisms and ultimately even to a brain which is capable of self-reflection.

Darwinian evolution is in fact a very powerful process, but it generally requires a self-replicating system which has some sort of a structural memory, for example, a genome, which is being copied from generation to generation. Several theories have been developed on how such a process may occur on a molecular level and how it can lead to molecular evolution. The most prominent example may be the RNA world [10] (see Chap. 6), but there are also alternative ideas like an interaction between RNA strands and peptides [11] (see also Chap. 8). In any case, it needs a relatively high degree of molecular development. So how could the basis for molecular evolution, the stage of self-replicating molecules, be reached?

In principle, there are two distinct pathways approaching the diagonal without Darwinian evolution (labelled as 1 and 2 in Fig. 9.3) [8]:

1. Systems which initially increase complexity and then gain order in a second step. Typically, a very large number of chain molecules with different sequences are formed in a random process (the state of high complexity) which then undergo rigid selection until a small number of "successful" chain molecules with distinct sequences are left (a state of reduced complexity but of higher order). A good example for this pathway 1 would be the formation and selection of peptides as described in Sect. 7.4.
2. Systems which initially increase order and then gain complexity in a second step. A typical example for this pathway 2 would be a process which starts on a mineral surface [12] (a state of high order) and then is getting transferred into a liquid or cellular environment (a state of reduced order but increased complexity).

So, there are in fact known molecular mechanisms which lead from simple, disordered states to complex and ordered states. It is now the challenge for the field of prebiotic chemistry to discover and identify them and to reproduce them experimentally.

Back to your role to search for life on a foreign planet: how could you recognize very early steps in the formation of life? This is a much more difficult task than just finding life, since you would not only have to look for states but you would have to look for slow developments. You would have to search for processes like the ones just mentioned (1 and 2) which lead into a pathway of increasing order and complexity. Actually, this is something one should try to do on earth as well: can we discover terrestrial processes which have this potential? If we identify such a process, we would indeed have come a large step closer toward understanding the true origin of life.

References

1. Benner SA (2010) Defining life. Astrobiology 10:021–1030
2. Schrödinger E (1992) What is life? Cambridge University Press, Cambridge

3. Stoddart MJ (2011) Mammalian cell viability: methods and protocols. Methods in molecular biology 740, Springer, New York

4. Johnson N (2010) Simply complexity: a clear guide to complexity theory. Oneworld Publications, London

5. Waters E et al (2003) The genome of Nanoarchaeum equitans: insights into early archaeal evolution and derived parasitism. Proc Natl Acad Sci U S A 100:12984–12988

6. Mushegian A (1999) The minimal genome concept. Curr Opin Genet Dev 9:709–714

7. Pennisi E (2001) The human genome. Science 291:1177–1180

8. Mayer C (2020) Life in the context of order and complexity. Life 10:5

9. Dell'Amore C (2013) Biggest viruses found, may be fourth domain of life? Natl Geogr 7:130718

10. Gilbert W (1986) The RNA world. Nature 319:618

11. Carter CW (2016) An alternative to the RNA world? Nat Hist 125:28–33

12. Wächtershäuser G (1988) Before enzymes and templates: theory of surface metabolism. Microbiol Rev 52:452–484

10

After LUCA: What Happened Next?

Abstract After rising to the earth's surface, the continued development of the first cells most likely took place in the vicinity of "black smokers" in the oceans. The flooding after meteor strikes also means that they could also have spread over the first continents. Here, independent strains of bacteria were able to develop which had no contact with other continents. With the onset of plate tectonics, a meeting of the various tribes came about following the collisions between continents which may have resulted in the endosymbiotic transfer of genes. The crust model presented also provides us with the chance to think about the role played by viruses entering into everyday life. One possibility is the parallel development of two similar RNA types, one of which remains inside the cells and one in the free environment inside the cavities.

10.1 The Triumphal March Begins

It is easy to imagine how constant eruptions by geysers continuously transported organic molecules to the surface of the earth along with cells after the formation of LUCA. In the vicinity of the exit points, "(bio) films" must have formed from not yet revitalized organic materials and later with the first cells that had little chance of surviving under the new environmental conditions. UV radiation, solar wind, lower temperatures, and higher pH values required an adjustment, which are certain to have taken a longer period of time. The supply of energy also caused difficulties. The adoption of organic molecules

© Springer Nature Switzerland AG 2020
U. C. Schreiber, C. Mayer, *The First Cell*, https://doi.org/10.1007/978-3-030-45381-7_10

from the existing biofilms probably only became possible following long steps in development. In the depth of the crust, other energy resources were used for the development. Sooner or later, climatic or regional conditions may have made it possible for the first cells on the earth's surface to survive, so that they could reach the oceans in flowing water. This raises the question of when the DNA replaced the RNA store as the main information store. One argument would be the higher stability of DNA in neutral to slightly alkaline environments, such as those found on the earth's surface. The extent to which this affects a feedback to cell chemistry is not clear, however. In addition to the adjustment due to stability, an expanded molecular system would also be needed, which made it possible to open, copy, and close a double-stranded system.

An alternative situation for the first cells rising to the surface is conceivable in locations near the coast, where the open fault systems led back directly into marine world. Their entering into the ocean probably defines the decisive step that set the triumphal march of the life system into motion. In the ocean, the cells encountered regions which they were very familiar with from the physicochemical conditions known to them. It was the black smokers and other hydrothermal vents which provided them with unlimited energy resources for use. At the same time, this environment again offered protection against the destructive influences present on the face of the earth.

Viewed in a manner compressed over time, the first cells taking over the submarine hydrothermal vents probably led to a first life form spreading like an explosion. The open ocean offered ideal conditions for this with its currents and hot springs in sufficient number. Long distances and unique local characteristics could lead to independent adaptations in the young family of cells, which quickly diversified as a result. This was the phase in which violent meteorite strikes were subsiding, which had mainly affected the oceans. Every major impact caused large areas of the young continents to be flooded. This was the way in which the prokaryotes were able to spread over all the large and small fragments of crust that had now been formed. Wherever they came into contact with hydrothermal vents, they could gain a foothold and continue developing. While 3.77 billion years ago already far-advanced bacteria presumably used the iron oxidation as an energy source [1], at least 3.5 billion years ago, bacteria (cyanobacteria) appeared for the first time which opened up an external energy source. The new invention they came forward with was a molecule that could convert the energy of sunlight into chemical energy. At the beginning, hydrogen and hydrogen sulfide were converted, but no oxygen was created as yet. The start of this was again linked to a molecule that had also undergone further development, chlorophyll. The beginning of oxygen production by cyanobacteria (formerly called blue-green algae) dates back to 2.8 billion years ago [2]. However, the oxygen content in the atmosphere only began to rise about 2.3 to 2.4 billion years ago, after a large proportion of the oxidizable components in certain minerals (Fe^{2+}, sulfur) reacted with oxygen. The increase in oxygen production was caused by the increased occurrence of stromatolites, algal mats in shallow marine areas, which have survived until today (Fig. 10.1). And as a consequence, this actually changed everything concerned with life.

Fig. 10.1   Recent stromatolites in Shark Bay, Western Australia, in saltwater with a high concentration of dissolved salts. The high salt content keeps predators away

10.2 The Contact Between Differently Developed Cells

It may seem far-fetched, but the keyword "crust fragments" inspires further considerations. Tectonic plates began moving on the earth at an early stage, which ultimately led to the displacement of the infant continents and subsequent collisions between continents. We are now certain that organelles, such as the mitochondria and chloroplasts from the eukaryotes, are derived from bacteria. We also believe that bacteria were assimilated by other prokaryotes but could not be digested. In fact, exactly the opposite is true: the ingested bacteria lived in the host cell and used its metabolic products as a supply source for themselves. A symbiosis developed in which one partner lived permanently in the body of the other, like the bacteria in our intestine. The evolutionary adaptation led to the enclosed bacteria also dividing in the same rhythm as host cells and gradually being reduced to useful building blocks. They make a decisive contribution to the supply of energy to cells today (endosymbiont theory) [3]. One of the consequences of this was that the cells were able to get bigger, thanks to an improved energy supply. But when could these events have taken place and how were they triggered?

In much later times, when a higher form of life was already fully developed, collisions between continents led to an intensive exchange of fauna and flora, which brought together the living environment which had previously been isolated. There was, for example, the great American fauna exchange 2.8 million years ago in relation, when North and South America close to what is now Panama were connected by a land bridge. The result was the drastic extinction of entire groups of animals while competing species flourished. What did a collision between continents mean at a time when only unicellular organisms existed but by being sufficiently isolated on each continent had developed completely differently? Were similar displacement effects in play or was this perhaps the reason why endosymbionts developed?

The eukaryotes appeared almost spontaneously on life's stage 1.5 billion years ago. Ancestors have not yet been clearly identified that could be viewed as transition stages. Five hundred million years previously, almost all existing small continents began to migrate toward each other, until 300 million years later a supercontinent had formed. The geologists call it Columbia [4]. It existed for another 200 to 300 million years, whereby the last 100 million years were already marked by decay and by mountain building processes caused by plate tectonics. The eukaryotes finally appeared in the final phase of

Columbia. One billion years earlier, the oxygen concentration in the atmosphere slowly began to increase, something the living world had to react to.

An exciting scenario can be built on this. It was assumed that on all distant small continents, endemic bacterial strains had developed over more than 1 billion years, which were gradually brought together on a large continent from a certain point in time within millions of years. Each time a new small continent landed at the large complex, different specialized bacterial species "suddenly" confronted each other. The number of individuals and time period were large enough for all variations of cohabitation to be tried out. It cannot be excluded that this process led to the development of the endosymbionts, perhaps even several times as a result of small continents subsequently arriving, as can be observed today in some cell organelles (plastids of brown algae, gold algae, and diatoms). If the adaptation to increasing oxygen concentrations in the atmosphere and ocean water took place at the same time, perhaps everything that came together when Columbia came to an end can explain how eukaryotes started.

We can link several questions to this hypothetical scenario here. Are there any relics from bacteria and archaea in the DNA today that point to a period of endemic development? Can perhaps provinces be found on the continents that correspond to old continent cores that are dominated by such bacteria, perhaps in the deep biosphere?

10.3   And the Viruses?

And we need to permit one more consideration. Every living being carries has a virus that is specifically tailored to its cells but which can infect and damage these cells. Only the constant race for evolutionary development with its constant adjustments on both sides has led to both the host and the viruses still existing side by side today. Viruses are actually particles that are not generally considered to be living beings. You could perhaps call them part-time creatures. They infect cells and can only multiply with the help of the cells. What is fully unclear is when they appeared in the world of living beings and how they developed in parallel with cells. Are they former bacteria, which having lost their ability to reproduce, therefore, need the support of other cells? Or are they strands of RNA and DNA that originated from their host and have now become independent? Another possibility is that they developed as part of a co-evolution from the first RNA building blocks in a separate line parallel to the cells. The model of vesicle formation presented in the continental crust offers new approaches to discussing the virus question. The parallel lines of the development of vesicles, proteins, and RNA in the cavities leave room for the simultaneous formation of RNA molecules, which continued to develop in the cells and which remained in the immediate vicinity but outside in the open system in the fault zones.

References

1. Dodd MS, Papineau D, Grenne T et al (2017) Evidence for early life in Earth's oldest hydrothermal vent precipitates. Nature 543:60–64
2. Olson JM (2006) Photosynthesis in the Archean Era. Photosynth Res 88(2):109–117. https://doi.org/10.1007/s11120-006-9040-5
3. Martin WF, Garg S, Zimorski V (2015) Endosymbiotic theories for eukaryote origin. Philos Trans R Soc Lond B Biol Sci 370(1678):20140330. https://doi.org/10.1098/rstb.2014.0330
4. Rogers JWS, Santosh M (2002) Configuration of Columbia, a Mesoproterozoic Supercontinent. Gondwana Res 5(1):5–22. https://doi.org/10.1016/S1342-937X(05)70883-2

Printed in the United States
by Baker & Taylor Publisher Services